P9-AEU-837

LEGITIMATE APPLICATIONS OF PEER-TO-PEER NETWORKS

LEGITIMATE APPLICATIONS OF PEER-TO-PEER NETWORKS

DINESH C. VERMA
IBM T. J. Watson Research Center

A JOHN WILEY & SONS, INC., PUBLICATION

This book is published on acid-free paper. ♾

For general information on our other products and services please contact our Customer Care Department within the U.S. at 877-762-2974, outside the U.S. at 317-572-3993 or fax 317-572-4002.

Wiley also publishes its books in a variety of electronic formats. Some content that appears in print, however, may not be available in electronic format.

Library of Congress Cataloging-in-Publication Data:

Verma, Dinesh, 1964-
Legitimate applications of peer to peer networks / Dinesh C. Verma.
p. cm.
Includes bibliographical references and index.
ISBN 0-471-46369-8 (Cloth)
1. Peer-to-peer architecture (Computer networks) 2. Application software—Development. 3. Electronic data processing Distributed processing. I. Title.
TK5105.525 .V37 2004
004.6'5—dc22 2003023099

Printed in the United States of America.

10 9 8 7 6 5 4 3 2 1

Dedicated to
Paridhi, Archit, and Riya

CONTENTS

PREFACE

Peer-to-peer computing has gained a lot of prominence and publicity in the past years. Several start-up companies offer software packages that can be used for building peer-to-peer overlay networks for applications such as file sharing and music swapping. However, the legal and ethical questions associated with swapping copyrighted material on the Internet have clouded the entire peer-to-peer technology industry and given peer-to-peer networks a bad name. Many companies and universities have banned peer-to-peer networks because of copyright infringement issues.

The vitriolic legal debate over copyright violations has the potential danger of eclipsing many different legitimate applications that can be developed over peer-to-peer networks. Such networks offer decentralized domains of control and an increased level of parallelism and are extremely resilient to failures of machines and network. As a result, peer-to-peer technology can be used to develop many new applications that would benefit from the unique features offered by the peer-to-peer architecture. Some companies already offer peer-to-peer software for applications such as instant messaging, search engines, and data storage.

The goal of this book is to describe some of the applications that can exploit the unique features of peer-to-peer networks and to discuss why developing these applications on a peer-to-peer network is better than developing them in the traditional manner. Peer-to-peer technology is not an appropriate choice for all applications, and the book also discusses the disadvantages that a peer-to-peer version of an application would have compared with a conventional client-server implementation of the same application.

WHO WILL BENEFIT FROM THIS BOOK?

This book is intended for enterprise IT operators, architects, software developers, and researchers who are interested in peer-to-peer networks. If you are a graduate student or network researcher who wants to learn about the different types of applications that could be developed on a peer-to-peer infrastructure, you will find this book to be very useful. The book contains a detailed description of many applications that can be developed on a peer-to-peer infrastructure and compares the relative merits and demerits of building those applications with a traditional client-server approach versus a peer-to-peer approach.

If you are a network operator who operates a corporate intranet, you will find the overview of peer-to-peer architecture contained in this book very helpful. This book will help you understand peer-to-peer systems and how you can monitor and control the peer-to-peer applications running on your intranet. You may want to pay special attention to the sections of Chapter 4 that describe the different measures and countermeasures to monitor and regulate your network bandwidth usage with peer-to-peer applications.

If you are an architect responsible for an enterprise network or an enterprise IT system and would like to harness peer-to-peer technology for use within your enterprise system, this book will help you understand the relative merits and demerits of peer-to-peer implementations of various applications compared with the traditional client-server implementations of the same application. This will help you decide among the different possible ways of implementing your own applications, as well as providing useful information to understand the trade-offs of different vendor products.

Finally, if you are a software developer working for a peer-to-peer software development company, or want to develop your own peer-to-peer application, this book will give you a comprehensive view of how different researchers have tried to approach problems similar to the ones you may be encountering in your specific application.

WHO IS THIS BOOK NOT FOR?

This book is not intended to provide information about the specific implementation of peer-to-peer networks. It does not describe the architecture of a specific peer-to-peer protocol but discusses general techniques that can be used for peer-to-peer network construction and the applications that can run on such networks. Although it provides a broad overview of many common peer-to-peer networks, you will not find it appropriate if you are looking for an in-depth treatment of a specific instance.

This book is not intended as a discussion of the legal issues related to running peer-to-peer systems. If you are looking for a legal discussion on running applications on peer-to-peer networks, this book is not for you. This is a technical book that describes various applications that can take advantage of peer-to-peer technology in various manners and is not intended for a legal audience.

ORGANIZATION OF THE BOOK

The material in this book is organized into 10 chapters.

Chapter 1 discusses the architecture of a peer-to-peer network and compares the peer-to-peer approach to traditional client-server architecture. It examines the key benefits of a peer-to-peer architecture and discusses its relative merits vis-à-vis the traditional client-server architecture.

Chapter 2 discusses how the different nodes in a peer-to-peer network can be connected together to create an overlay network connecting the participants over the Internet. It also discusses how the nodes discover each other in order to create such an overlay.

Chapter 3 discusses a common building block that supports many of the applications that are run on peer-to-peer systems, namely, the ability to multicast messages to a number of participants without relying on any capabilities of the underlying network. This type of multicast communication performed at the application layer is essential for the operation of most peer-to-peer applications.

Chapter 4 provides a brief overview of the ubiquitous application driving many peer-to-peer systems: the sharing of files on the peer-to-peer network. It discusses the technology used for such applications and describes the different techniques that can be used to improve the efficiency and security of file-sharing applications. It also discusses the legal and resource usage issues that arise when peer-to-peer networks are used to illegally exchange copyrighted material. It describes the measures enterprise operators, ISPs, and copyright holders can use to prevent illegal exchange on peer-to-peer networks as well as the countermeasures that peer-to-peer developers can use to work around those controls.

Chapter 5 discusses the use of peer-to-peer networks to build file storage services on the Internet. Peer-to-peer computing can be used to provide storage systems that are capable of providing ubiquitous access to files of a user from any location. The techniques and mechanisms needed to build file storage systems are described in this chapter.

Chapter 6 discusses how peer-to-peer networks can be used to build a data backup service without requiring expensive backup servers or a large amount of disk-space. The assumptions under which such a service works well are also discussed in this chapter.

Chapter 7 discusses how a directory service can be improved by using a peer-to-peer infrastructure. The directory service on peer-to-peer infrastructure can provide more scalable solutions when distributed administration domains are involved in managing different portions of the directory.

Chapter 8 discusses how peer-to-peer systems can be used to build publish-subscribe middleware, which is commonly used in many enterprises to create different types of distributed applications. The advantages and disadvantages of a peer-to-peer implementation versus a traditional client-server implementation are discussed.

Chapter 9 discusses a variety of different applications that can be built on top of the different middleware and systems that are described in the earlier chapters. It includes examples of peer-to-peer instant messaging, peer-to-peer IP telephony, collaborative databases, collaborative hosting of content, and anonymous web surfing, all applications that can be built in a scalable manner with peer-to-peer technology.

Finally, Chapter 10 discusses some of the topics that are related to peer-to-peer technology. It includes a discussion of peer-to-peer applications that were developed before file sharing propelled peer-to-peer technology into the evening news as well as grid computing, which has many similarities with peer-to-peer systems.

1

THE PEER-TO-PEER ARCHITECTURE

In this chapter, we look at the general architecture of a peer-to-peer system and contrast it with the traditional client-server architecture that is ubiquitous in current computing systems. We then compare the relative merits and demerits of each of these approaches toward building a distributed system.

We begin the chapter with a discussion of the client-server and peer-to-peer computing architectures. The subsequent subsections look at the base components that go into making a peer-to-peer application, finally concluding with a section that compares the relative strengths and weaknesses of the two approaches.

1.1 DISTRIBUTED APPLICATIONS

A distributed application is an application that contains two or more software modules that are located on different computers. The software modules interact with each other over a communication network connecting the different computers.

To build a distributed application, you would need to decide how many software modules to include in the application, how to place those software modules on the different computers in the network, and how each software module discovers the other modules it needs to communicate with. There are many other tasks

Legitimate Applications of Peer-to-Peer Networks, by Dinesh C. Verma
ISBN 0-471-46369-8

that must be done to build a distributed application, but those mentioned above are the key tasks to explain the difference between client-server computing and peer-to-peer computing.

1.1.1 A Distributed Computing Example

The different approaches to distributed computing can be explained best by means of an example. Suppose you are given the task of creating a simulation of the movement of the Sun, the Earth, and the Moon by a team of five astronomers. Each of the five astronomers has a computer on which he or she would like to see the motion and position of the three heavenly bodies at any given time for the last 2000 years as well as the next 2000 years. Let us say (purely for the sake of illustration, rather than as the preferred way to write such simulation) that the best way to solve this problem is to create a large database of records, each record containing the relative positions of the three bodies at different times ranging over the entire 4000-year period. To show the positions of the three heavenly bodies, the program will find the appropriate set of records and display the position visually on the computer screen. Even after making this choice on how to write the program, you, as the programmer assigned to the task, have multiple ways to develop and deploy the software.

You can write a stand-alone program that will do the complete simulation of the three heavenly bodies that runs on a single computer and install five copies of it on each of the computers. This approach (approach I) has the advantage that each astronomer can run the program as long as his or her computer is up and does not require access to a network, or to the computers of the other astronomers. This approach would be fine if the application runs well enough on each of the computers, but it does not harness the combined processing power of all the five computers. Furthermore, experienced programmers know that all programs must be maintained and upgraded multiple times—to fix bugs, to add new features, or to correct any errors in generation of the set of records in the program. With this approach, any changes that you make after the initial installation of the program would need to be replicated five times.

An alternative approach (approach II) would be for you to select the most powerful computer among the five to do the simulation, with all the other computers (as well as the one running the simulation) having a visualization interface for the users to interact with the simulator. You have broken the program into two software modules, the visualization module and the simulation module, and created five instances of the visualization module and one instance of the simulation module. If the astronomers' computers are of differing power, this allows all of them to harness the power of the fastest computer. Because the simulation module maintains a large set of data, this set can be maintained at a single place and use less disk space. Also, you can localize changes to the simulation module to a single computer, and only changes to the visualization module must be propagated to all of the five computers. If the visualization module is much simpler than the simulation module, this will cut down significantly on the number of bugs and changes that need to be maintained in different places. The drawback now is that each astronomer needs connectivity to the network in order to access the simulation module and that the computer running the simulation module must be available continuously.

Approach II outlined above follows the client-server architecture for distributed applications where the fastest computer is acting as the simulation server. The visualization modules running on the other computers are the clients that accesses the simulation module running on the server.

Although approach II allowed each computer to access the resources of the fastest computer, it did not use the combined processing power of the other four computers available to the distributed application. To use the processing power of all the five computers, you can divide your simulation modules into five identical portions, each one handling a different but similar part of the simulation process (approach III). Recalling the fact that the simulation module was implemented as a database of relative positions, each of the five computers can be assigned to hold a portion of the database. One could split the database into five equal portions, each computer holding one portion, or one could divide the database into overlapping portions so that the position at any time is stored at two or more computers. If disk space is not an issue, one could simply replicate the database on all the computers. When an astronomer wants to check the position of the three heavenly bodies at any time, the visualization module on his/her computer finds one of the five computers that has the simulation module with the correct portion of the database and then talks to that simulation module. All the five computers are acting as peers, each having a client component (the visualization module) as well as a server component (the simulation module). Approach III is the pure peer-to-peer approach to solving the three-body simulation problem.

Approach III could potentially be more scalable than approach II because it is leveraging the combined power of all the computers rather than that of a single computer. If the records in the database are available from multiple computers, the reliability of this approach may be higher than that of approach II. However, it reintroduces the problem that any changes made to a module (simulation or visualization) need to be replicated on all of the different computers.

In real life, one could also use a hybrid approach that is a mixture between the client-server architecture and the peer-to-peer architecture. The hybrid approach places some software modules on a set of computers that can act as servers and others act as clients. The hybrid approach for some distributed applications can often result in a better trade-off between the ease of software maintenance, scalability, and reliability.

For any of the approaches selected, you would need to solve the discovery problem. The different modules of the application need to communicate with each other, and a prerequisite for this would be that the modules know where to send messages to the other modules. In the Internet, messages are sent to other applications by specifying their network address, which consists of the IP address of the application and the port numbers on which the application is receiving messages. To communicate over the Internet protocol suite, each software module must find out the network address of the other software module (or modules).

One solution to the discovery process is to fix the port numbers for all the software modules that they will be using and have all the modules know the port numbers and IP addresses of the different modules. When developing the simulation application for the astronomers, you can hard code this information within each of the modules. However, you must ensure that the selected port numbers are available on the computers that the applications will be running on. Because most computers run applications developed by many different companies, this solution would require a global coordination of port numbers among all the software developers in the world, which is clearly not feasible. The alternative is to have the address and port number information be provided as configuration parameters to the different software modules. If you use this approach with the example application we have discussed here, it is relatively easy to specify five port numbers and IP addresses in the configuration of each computer. However, if you consider the case of a more complex real-world application that needs to run on many more computers, the manual effort required for configuration could be quite substantial.

One of the key advantages of the client-server architecture (approach II discussed above) is that it makes the discovery process quite simple. This enables the deployment of a large number of clients and a high degree of scalability. Let us now define the client-server architecture and the peer-to-peer computing architecture in a more precise manner and then examine the discovery process in each of the architectures.

1.1.2 Client-Server Architecture

The client-server architecture is a way to structure a distributed application so that it consists of two distinct software modules:

- A server module, only one instance of which is present in the system
- A client module, of which multiple instances are present in the system

The only communication in the system is between the client modules and the server module.

Please note that the client and server modules themselves may be quite complex systems with further submodules and components. However, the key characteristic of the client-server architecture is that there is a server module that is the central point for communication. Clients do not communicate with each other, only with the server module.

In the client-server architecture, the server is usually the more complex piece of the software. The clients are often (although not always) simpler. With the wide availability of a web browser on most desktops, it is quite common to develop distributed applications so that they can use a standard web browser as the client. In this case, no effort is needed to develop or maintain the client (or, rather, the effort has been taken over by a third party—the developer of the web browser). This simplifies the task of maintaining and upgrading the application software.

In any distributed application, the different components must discover each other in order to communicate. In the client-server architecture, only the clients need to communicate with the server. Therefore, each client needs to discover the network address of the server, and the server needs to know the network address of each of the clients.

The solution used for discovery in the client-server architecture is quite simple. The server runs on a port and network address that is known to the client module. The clients connect to the server on this well-known network address. Once the client connects to the server, the client and server are able to communicate with each other. The server need not be configured with any information about the clients. This implies that the same server module can communicate with any number of clients, constrained only by the physical resources needed to provide a reasonable response time to all of the connected clients.

For most common applications that run on the Internet, the port numbers on which the server side can run have been standardized [1]. Thus the clients only need to know the IP address of the computer on which the server is running. Any individual client can also easily switch to another server module by using the IP address (or, in general, the IP address and the port number) of the new server. As an example, a web server typically runs on port 80 and web browsers can connect to the web server when a user specifies the name of the computer running the web server. The browser also has the option of connecting to a server running on a port different than 80.

The simplicity and ease of maintenance of client-server architecture are the key reasons for its widespread usage in the design of distributed applications at the present time. However, the client-server architecture has one drawback—It does not utilize the computing power of the computers running the client modules as effectively as it does the computing power of the server module. At present, when even the standard desktop packs more computing power than the computers that were used for Neil Armstrong's flight to the Moon in 1969, this does appear to be a rather wasteful approach.

1.1.3 Peer-to-Peer Architecture

The peer-to-peer architecture is a way to structure a distributed application so that it consists of many identical software modules, each module running on a different computer. The different software modules communicate with each other to complete the processing required for the completion of the distributed application.

One could view the peer-to-peer architecture as placing a server module as well as a client module on each computer. Thus each computer can access services from the software modules on another computer, as well as providing services to the other computer. However, it also implies that the discovery process in the peer-to-peer architecture is much more complicated than that of the client-server architecture. Each computer would need to know the network addresses of the other computers running the distributed application, or at least of that subset of computers with which it may need to communicate. Furthermore, propagating changes to the different software modules on all the different computers would also be much harder. However, the combined processing power of several large computers could easily surpass the processing power available from even the best single computer, and the peer-to-peer architecture could thus result in much more scalable applications.

The bulk of this book is devoted to the subject of peer-to-peer applications. In Section 1.2, we look at the architecture of the typical software that must run on each computer in the peer-to-peer architecture. Subsequent chapters in the book discuss the issues of discovery and creating communication overlays among all the nodes that are participating in the peer-to-peer architecture.

1.2 THE PEER-TO-PEER SOFTWARE STRUCTURE

As mentioned above, a distributed application implemented in a peer-to-peer fashion would have the same software module running on all of the participating modules. Given the complexity associated with discovering, communicating, and managing the large number of computers involved in a distributed system, the software module is typically structured in a layered manner. The software of most peer-to-peer applications can be divided into the three layers, the base overlay layer, the middleware layer, and the application layer.

The base overlay layer deals with the issue of discovering other participants in the peer-to-peer system and creating a mechanism for all the nodes to communicate with each other. This layer is responsible for ensuring that all the participants in the nodes are aware of the other participants. Functions provided by the base layer are the minimum functionality required for a peer-to-peer application to run.

The middleware layer includes additional software components that could be potentially reused by many different applications. The term "middleware" is used to refer to software components that are primarily invoked by other software components and used as a supporting infrastructure to build other applications. Functions included in this layer include the ability to create a distributed index for information in the system, providing a publish-subscribe facility, and security services. The functions provided in the middleware level are not necessary for all applications, but they are developed to be reused by more than one application.

Finally, the application layer provides software packages intended to be used by human users and developed so as to exploit the distributed nature of the peer-to-peer infrastructure.

Some implementation of the base functionality is needed in all peer-to-peer systems. Some peer-to-peer systems [2, 3] only provide middleware functionality and enable other applications to be written on top of them. Other peer-to-peer systems provide complete applications by using a common middleware [4] or by providing their own private implementation [5].

Please note that there is no standard terminology across different implementations of peer-to-peer systems, and thus the terms used above are general descriptions of the functionality needed for building a generic peer-to-peer system, rather than the structure of any specific peer-to-peer system. Also, most peer-to-peer systems are developed as single applications. However, the structuring of the three layers as discussed in this chapter provides a good way to categorize and study the different applications.

1.2.1 Base Overlay Layer

As mentioned above, the base overlay formation is a feature that must be provided by all peer-to-peer systems. The functions included in this layer include the following:

- *Discovery:* Before communicating with each other, a node in a peer-to-peer system must discover a minimum set of other nodes so that it could communicate with them. The discovery mechanism may include discovering all the other nodes in the system or just one other node that could be used as an intermediary to communicate with the other nodes.
- *Overlay Formation:* This provides a mechanism by which all the peer-to-peer nodes are connected into some type of common network. The network is used by each of the nodes to communicate with the other nodes.
- *Application-Level Multicast:* This functionality permits a node to send a message out to all of the other nodes in the network. In some peer-to-peer infrastructures, the only communication supported is the ability of a node to send a message to all of the other nodes. Some other peer-to-peer architectures would allow formation of subgroups within the system so that the message is sent only to a restricted subset of nodes.

1.2.2 Middleware Functions

The middleware layer is responsible for providing some common functions that will be used by applications at the higher layer. The middleware consists of those software functions that are intended to be used primarily by other software components, rather than by a human user. The middleware function in itself cannot be used to build a complete application, but the common functions can be used to build peer-to-peer applications rapidly.

Some of the functions included in this layer are:

- *Security:* This middleware function provides the support needed for managing secure communication among the different nodes, providing support such as encryption, access control, and authentication. The issues involved in the security aspects of peer-to-peer communication are similar to those involved in traditional client-server computing systems. However, the distributed nature of the system makes security issues much harder in peer-to-peer systems.
- *Distributed Indexing:* Many peer-to-peer applications need a fast way to index and find information that is distributed along the different nodes of a peer-to-peer infrastructure. A

distributed indexing system could be used by applications such as a distributed storage application or a distributed file system. A special type of an index is a hash table, which maps keywords of arbitrary length into a fixed-length hash and uses the hash to locate the entries corresponding to the keywords. Distributed indexing and hash tables have been an active area of research in peer-to-peer computing systems.

- *Directory Services:* A directory service provides a name lookup service, whereby one could look up the properties of an entity by specifying its name. In many respects, a directory service is like a index or hashing service, with one difference. It is common in conventional directory services to impose a hierarchical naming structure on the elements stored in the directory. The most widespread directory servers use the LDAP protocol to allow clients to access the information stored in the directory service. A peer-to-peer implementation of the directory service can offer some unique advantages over the traditional server-based implementation.
- *Publish-Subscribe Systems:* A publish-subscribe system allows for the sharing of information in a system in a controlled manner. Publishers of information send the information to the publish-subscribe system, and the subscribers of information inform the system about what types of information they wish to receive. The publish-subscribe system manages the published information sources and the preferences of the different subscribers and provides an efficient delivery mechanism.

Most middleware functions can be implemented with a client-server approach as well as a peer-to-peer approach. In the subsequent chapters of the book, we look at the different middleware functions that can be provided with a peer-to-peer infrastructure and compare the alternative implementations of the middleware functions.

1.2.3 Application Layer

We define this layer as consisting of the software components that are designed to be used primarily by a human user. The file-sharing application is the most ubiquitous peer-to-peer application, with multiple implementations available from a large number of providers. The file-sharing application allows users of a peer-to-peer network to find files of interest from other computers on the network and to download them locally. The use of this application for sharing copyrighted content has been the subject of several legal cases between developers of peer-to-peer software and the music industry.

File sharing, however, is not the only application that can exploit the properties of a distributed base overlay infrastructure. A peer-to-peer infrastructure can be used to support self-managing websites, assist users to surf the network in an anonymous manner, and provide highly scalable instant messaging services and a host of other common applications.

There are some old applications which were built and developed with the peer-to-peer model long before the file-sharing application grew in prominence. These applications include some routing protocols used within the Internet infrastructure as well as the programs used to provide discussion and distributed newsgroups on the Internet.

Several such applications, new and legacy, are discussed in the subsequent chapters of this book. Each of these applications can be implemented with the client-server architecture or the peer-to-peer architecture. With each application, there are unique advantages and drawbacks in development using the peer-to-peer structure as compared to the client-server structure. However, some of these advantages and disadvantages pertain across all such applications and are discussed in Section 1.3.

1.3 COMPARISON OF ARCHITECTURES

If you had to implement an application and had the choice of implementing it with a peer-to-peer architecture or a client-server architecture, which one would you pick? Either of the two approaches to building the application can be made to work in most cases. In this section, we look at some of the issues you should consider when deciding which of the two approaches would be more appropriate for the task at hand. Each of the subsections discuss some of the issues you may want to consider and the merits and demerits of the two architectures compared with each other.

1.3.1 Ease of Development

When building an application, you need to consider how easy or difficult it will be to build and test the software for the application. The task of developing the software is eased by the existence of development and debugging tools that can be used to hasten the task of developing the application.

For developing client-server applications, there are a large number of application programs that are available to ease the task of development. Many software components, such as web servers, web-application servers, and messaging software, are available from several vendors and provide infrastructure that can be readily used to provide a server-centric solution.

Some programming environment packages are available for peer-to-peer computing, such as Sun's JXTA package [6] or Windows XP P2P SDK [7]. However, these packages are relatively new compared with the more traditional client-server software and therefore not quite as mature. Thus the risk of running into an undiscovered bug in the infrastructure is higher for peer-to-peer packages compared with those for client-server computing.

Furthermore, the task of debugging and testing a centralized server solution is easier than the task of debugging distributed software that requires interaction among several components. Thus, from an ease of development perspective, the client-server approach has an advantage over the peer-to-peer approach.

1.3.2 Manageability

Manageability refers to the ease of managing the final software when it is finally deployed. After a software application is up and ready, it still needs ongoing maintenance while in operation. Maintenance includes tasks such as ensuring that the application has not stopped working (and restarting it in case it stops working), making backup copies of the data generated by the application, applying software upgrades, fixing any bugs that are discovered, educating users about the application, and a variety of other functions.

Several tasks (e.g., user education) associated with managing a running application remain essentially unchanged regardless of the implementation approach that is used. However, many tasks (e.g., backup, upgrades, bug fixes) are easier to do on a centralized application than on an application that is run on multiple platforms. An associated issue with peer-to-peer applications is the different number of platforms that the software would need to run on. For a centralized client-server approach, the server part of the software is at a single location and one could choose the platform on which the server part of the application would run. Choosing a platform in this instance means selecting the hardware and operating system of the machines on which the application software will run. A common platform allows for an improved degree of manageability on the server component of the software. The platform choices for the client component cannot be similarly restricted. However, for a client-server architecture that is developed with a standard client (e.g., a web browser), very little maintenance is associated with the client portion.

In a peer-to-peer application, the application is running on several different machines that could be distributed across a wide geographic area. If they are all under a single administrative domain, it is possible to standardize on a common platform for all of them. However, it is more common to find the situation in which the different components of a peer-to-peer application would run on different platforms. This makes the manageability of peer-to-peer applications much harder.

The advent of platform-independent programming languages such as Java has simplified the manageability of distributed peer-to-peer applications to a large extent. Furthermore, the dominance of the Windows computing platform on the desktop market has limited the number of potential platforms that a peer-to-peer application needs to run on. These factors help the manageability of peer-to-peer applications to a large extent. However, in general, a peer-to-peer application is typically less manageable than a client-server application.

1.3.3 Scalability

The scalability of an application is measured in terms of the highest rate or size of user-level interactions that the application can support with a reasonable performance. The quantity in which scalability is measured is determined by the type of application. The scalability of a web server can be measured by the number of user requests it can support per second; the scalability of a directory server can be measured by the number of operations it can support per second, as well as by the maximum number of records it can store while maintaining a reasonable performance, for example, the maximum number of records it can store while keeping the time for a lookup operation below 100 ms.

Peer-to-peer applications use many computers to solve a problem and thus are likely to provide a more scalable solution than a server-centric solution, which relies on a single computer to perform the equivalent task. In general, using multiple computers would tend to improve the scalability of the application compared with using only a single computer.

However, a server-centric solution could be developed that uses multiple computers as well. Most high-performance server sites typically deploy many computers with a load balancer or dispatcher in front of the servers to provide a scalable solution. A dispatcher device distributes incoming requests to one of the many servers. Each server can process the request in the same manner. In most cases, the dispatcher maintains the affinity between clients and servers, that is, the dispatcher remembers which client requests were forwarded to which server and forwards multiple requests from the same client to the same server. An increase in the number of servers helps the system handle a large volume of requests. More details can be found in [8].

In general, if one uses the same number of computers to solve the problem, using all of the computers as servers would provide a more scalable system than using the same number of computers in a distributed peer-to-peer manner. This is because the peer-to-peer infrastructure requires communication among different nodes to perform the various tasks and thus has a larger overhead compared with a server-centric approach. In most cases, a centralized solution is much more efficient than a distributed solution.

The scalability of client-server computing as well as peer-to-peer computing has been proven by experience. Web servers of popular websites such as cnn.com or yahoo.com can handle millions of requests each day on a routine basis. Similarly, the large number of files exchanged on existing peer-to-peer networks such as gnutella and kazaa is on the order of millions of files every day. However, there is one difference between the scalability of the centralized server solution and the peer-to-peer solution. To build a server-centric solution, one would typically need to procure dedicated computers and host them at a facility. It is possible in many peer-to-peer applications to leverage existing computers (e.g., workstations and laptops) that are not always heavily used. This enables a peer-to-peer infrastructure to harness many more computers at very little cost and thus allows a more scalable solution at a lower cost.

A fair criticism of the low-cost aspect of peer-to-peer systems such as gnutella or kazaa is that they are getting a free ride on the costly infrastructure paid for and supported by users and Internet service providers (ISPs). Although the criticism is valid, nothing prevents enterprise or industry operators to use the same trick to provide low-cost solutions to scalability. Most enterprises have hundreds of computers, desktops, and laptops, which are only using a fraction of their computational power and capacity. By harnessing their computing power with a peer-to-peer approach, enterprises can build scalable applications at a lower cost than that of comparable systems using dedicated servers.

1.3.4 Administrative Domains

One of the key factors determining how to structure the application would depend on the usage pattern of the application and how the different computers that are used to deploy the application software are going to be administered. In general, with a client-server approach, the server computers need to be under a single administrative domain. Thus the server-centric approach would typically not be used if the servers needed to host the application would belong to several different administrative domains.

A peer-to-peer system, however, can often be created by using computers from many different administrative domains. Thus, if usage of the software requires that computers from many different administrative domains be used, the peer-to-peer approach would be the natural choice for that application.

1.3.5 Security

Once an application has been deployed, one of the administrative tasks associated with it is to manage its security. Security management entails the tasks of making sure that the system is only accessed by authorized users, that user credentials are authenticated, and that malicious users of the system do not plant viruses or Trojan horses on the system.

Security issues and vulnerabilities have been studied comprehensively in server-centric solutions, and safeguards against the most common types of security attacks have been developed. In general, the security of a centralized system can be managed much more readily than the security of a distributed infrastructure. In a distributed infrastructure, the security apparatus and mechanisms must be replicated at multiple sites as opposed to a single site. This increases the cost of providing the security infrastructure. Furthermore, the existence of multiple sites allows for an increased vulnerability because there are more points that can be attacked by a hacker. In peer-to-peer applications that are written to run on computers across multiple administrative domains, the security issues are even harder to solve.

1.3.6 Reliability

The reliability of a system is measured by its ability to continue working when one or more of its components fail. In the context of computer systems, reliability is often measured in terms of the ability of the system to run when one or more computers hosting the system are brought down. The approach used for reliability in most computers is to provide for redundant components, having multiple computers do the task instead of a single computer, such as having standbys that can be activated when a primary computer fails.

High-reliability computer applications can be developed by using either client-server or peer-to-peer architectures. The solution for scalability for high-volume servers also provides for increased reliability and continued operation in case one of the servers fails. Distributed peer-to-peer systems, for most applications, use multiple computers to do identical tasks, and thus the system continues to be operational and available, even when a single computer fails or goes off-line. The most popular peer-to-peer networks are made up of thousands of computers. Although each computer in itself is a simple desktop and goes out of service frequently (when users switch off their machines), the entire system keeps on functioning without interruption.

As in the case of scalability, the difference between the reliability of a server-centric approach and the peer-to-peer approach is that of the cost at which the reliability is achieved. The peer-to-peer approach provides for a much more lower-cost solution for reliability than the server-centric approach.

Summarizing the overall discussion, we can say that a client-server approach provides for better security, manageability, and ease of development, whereas the peer-to-peer approach provides for increased reliability and scalability in a more cost-efficient manner and allows for interoperation across multiple administrative domains.

2

PEER DISCOVERY AND OVERLAY FORMATION

A peer-to-peer system consists of multiple participants that are located in a network. The network could the global Internet, the intranet contained within an enterprise, a local area network, or even a home network connecting the different peers. The same peer-to-peer software is located on all of the participants. When the software on any given participant is started, it must connect to the other participants as the first step. To do so, it first must find who the other participants in the system are and how to connect to them. The process of finding other participants is the discovery step, and the process of interconnecting with the other participants is the overlay formation step.

In this chapter, we look at the different techniques that can be used for the discovery and overlay creation steps. We assume that the environment consists of a set of participants that all desire to become part of the overlay. We assume that the participants are all connected by a network, so that when a minimum set of attributes about a participant (e.g., its Internet address and port number) are known to another participant, the two are able to communicate.

Legitimate Applications of Peer-to-Peer Networks, by Dinesh C. Verma
ISBN 0-471-46369-8

2.1 DISCOVERY

One of the participants that will be part of the peer-to-peer system is the first one to come up and would be the first active one in the system. The first participant has to wait for other participants in the system to come up and join the system. The first participant could passively wait for the subsequent participants to contact it. The subsequent participants that come up in the system must find the first participant, or one of the other participants that have already come up previously, and connect to it. Most peer-to-peer systems have been designed so that the new participants joining the overlay connect to existing participants. However, it is also possible in some special cases for the participants in the existing overlay to try to find potential new participants proactively.

The discovery process tells a new participant about the other participants in the system. When a participant discovers another participant, it finds about the properties of that participant. The minimum property that is discovered is the identity of the other participant. However, other properties of the participant would also need to be discovered depending on the specific implementation of a peer-to-peer network. As an example, in a peer-to-peer implementation that uses a fixed port for communication, the discovery process only needs to find out the Internet address of another peer to communicate with it. Peer-to-peer implementations that can use any available port on the system must discover the Internet address as well as the port number being used by the other peer. Other systems may want to include properties that are specific to the application, for example, the type of machine that one has, as well as the operating system of the machine. However, once the minimum set of properties needed to facilitate communication between participants is discovered, two participants can discover other properties by communicating with each other.

The following subsections describe some of the common techniques used for discovery in peer-to-peer networks.

2.1.1 Static Configuration

The simplest process for discovery is to configure the peer-to-peer software on all the participating participants so that they contain the Internet address and/or port of the other peers in the system. When the peer-to-peer software starts on any machine, it checks the participants listed in the configuration to see if they are up and connects to them. A participant can also check periodically on the participants that are listed in the configuration but have not yet connected to see whether they subsequently come up.

If the set of participants that will participate in the peer-to-peer system is fixed, then defining the list of participants in the configuration information is the only task that needs to be done to complete the discovery process. For peer-to-peer systems in which the participating participants remain in the system for the long haul and the number of participants is moderate, static configuration is a viable approach. If the number of participants in the system is large, the configuration information would become huge and the configuration step will be tedious to do manually. Similarly, if the set of participants that are in the system keeps on changing frequently, for example, participants keep on leaving and joining the system, static configuration will not be appropriate for the task at hand. The centralized directory approach provides a better solution in these cases.

2.1.2 Centralized Directory

When a centralized directory approach is used for discovery, the configuration of the peer-to-peer software on each participant contains the name of a directory server. The directory server is a special participant that contains the listing of all the participants that are participating in the system. The directory server can contain information about the network address of the participants, the ports on each participant that are used by the peer-to-peer application, and also, optionally, other information such as the application running on the participants and even application-specific data, for example, a listing of the files that are available at a participant. To discover the other participants in the system, each participant simply queries the directory server to obtain this information.

When the central directory server approach is used, the peer-to-peer software on each participant in the system can dynamically choose the port on which the software runs and then register its IP address and selected port to the directory server. It can also find out the ports being used by other peers in the system, as well as obtaining a list of other participating peers in the system. When the peer-to-peer software on the participant terminates, it sends a message to the directory server requesting that it be removed from the list of participants. If each participant in the system registers and removes itself from the system as appropriate, the information in the directory server will always be consistent and up-to-date with the current set of participants.

Unfortunately, some instances of peer-to-peer software running on a participant may be terminated abruptly and may not always be able to send a termination message. This can happen when the machine on which the software is running crashes or the connectivity of the machine to the network is disrupted unexpectedly. A typical solution to this problem is to have each participant reregister itself periodically with the directory server. If a participant has not reregistered after a few periods, it is removed from the list of active participants. This keeps the information in the directory server reasonably up-to-date. As an example, let us assume that the participants register themselves every 10 minutes with the directory server and the directory server removes any participant that has not registered itself within the last 30 minutes. The directory server has a current set of active participants that contains all participants that have not encountered abrupt disruption within the last 30 minutes. There is an obvious trade-off between the rate of registration requests arriving at the directory server and the staleness of the information that is maintained at the directory server.

In most peer-to-peer systems, a new participant trying to join the system does not need to know all the participants that are present in the system. As long as the new participant is able to get a list of a few participants that are in the system, it will be able to join the overlay. Thus the fact that the directory server has a few extra entries, participants that are listed in the system as being present but have disappeared, is not a significant problem. As long as most of the participants listed in the directory server are currently within the peer-to-peer system, the directory server can provide a list of existing participants within the system that is current enough to allow the new participant to join the peer-to-peer system.

As an example, consider the directory server with 1000 participants listed. Let us say that 10% of these participants are listed incorrectly within the directory server—they have actually been stopped and were unable to inform the directory server. Let us assume that a new participant needs to learn about one existing participant in the system in order to join the peer-to-peer system. If the directory server provides the new participant with a list of six participants that it has currently registered with itself, then the new participant can join the peer-to-peer system as long as any one of those participants is still in the system. The chance that all six participants are not in the system, and therefore the new participant will not be able to join, will be less than one in a million. Thus, even with an unrealistically large percentage of inaccurate information, the directory server scheme can provide information to new participants that will have a very high probability of succeeding.

The directory server approach to discovery is efficient and is capable of handling a large number of participants that are dynamically joining and leaving the peer-to-peer system. Its scalability is only limited by the number of participants that the directory server is capable of handling. The drawback of the directory server is that it provides a single point of failure. If the directory server fails, new participants will not be able to join the system and other system functions that depend on the presence of the directory server will be adversely affected. The directory server also provides a vulnerability point that can be targeted by hackers to disrupt the function of the peer-to-peer system.

2.1.3 Using the Domain Name Service

The current Internet already has a large-scale distributed directory service that is actively used by all the computers connected to the Internet. The domain name service (DNS) is used by all computers on the Internet to resolve human-readable host names to the corresponding IP address. Thus the domain name address of the server www.wiley.com maps to the Internet address of 208.215.179.146. The DNS answers queries from most systems needed to provide the name translation service. The DNS is designed so that multiple IP addresses can be returned for the same domain name address. For sites that need to support a large volume of user requests, it is common to deploy multiple computers to handle the volume of requests. If those computers are assigned different IP addresses, the DNS could return the addresses of those multiple computers in response.

This feature of the DNS can be exploited to provide the discovery of other members in a peer-to-peer network. If a domain name is registered for the users of a peer-to-peer network, any member can query the DNS to obtain the list of all (or some) of the other members that are participating in the DNS. The advantage of this approach is that it exploits a stable and longstanding service of the Internet. The disadvantage of the approach lies in the fact that each new member must register its IP address with the DNS when it starts up. If the number of members is very large (e.g., typical peer-to-peer systems may have thousands of members), the DNS infrastructure would not be able to handle such a large number of addresses for a single domain name. Furthermore, security issues would arise when arbitrary computers are allowed to register their IP addresses as aliases for an existing domain name.

There are ways to exploit the DNS to work around the problems mentioned above. The DNS allows individual organizations to run their own servers, which are responsible for managing the names assigned within their own system. Thus all computer domains names ending in ibm.com would be administered by IBM Corporation and be registered in a DNS operated by the IBM Corporation. A directory service could be run by the operator of a peer-to-peer network, and the registration protocol between the participants and the directory service could be modified to address security concerns. The directory service could hand out a limited set of numbers of current participants on a rotating basis. The member propagation techniques described in Section 2.1.4 can then be used by participants to discover other members that are participating in the peer-to-peer system.

2.1.4 Member Propagation Techniques with Initial Member Discovery

In general, a participant in a peer-to-peer network does not need to discover all of the participating members in the network. In many cases, it is adequate to discover only a subset of the participating members, that is, only a smaller set of participants that can allow it to provide sufficient connectivity to participate in the peer-to-peer system. Even when a peer-to-peer participant needs to learn about all the members, it need not learn about all of them at once. In most cases, it is adequate for a new member to discover a small number of existing members (usually one is adequate) and then to learn about the remaining members from them.

2.1.4.1 Member Propagation with Full Member List. Let us consider the case when a new member needs to learn the list of all the members that are participating in the peer-to-peer system. The number of all the members needed implies that the overall size of the peer-to-peer group for such an application would be relatively small. However, one side-benefit of this approach is that each of the existing members in the peer-to-peer group would have information about all of the other participants that are in the existing group. As a result, a new member that is joining the group can obtain the full list of members as long as it is able to contact one of the existing members in the system. The various techniques mentioned previously in this chapter provide a way for the new participant to find one such existing member. The full list of participants can be loaded from the member thus discovered.

After a new member joins, the list of members maintained at each of the participants must be updated with the information about the new member. Once connected, the new member can inform the peer it is connected to to forward its information to all of the other participants. The broadcast mechanisms described in Chapter 4 provide multiple ways of propagating the information to other participants.

2.1.4.2 Member Propagation with Partial Member List. Maintaining the full member list has several drawbacks that limit the scalability of a peer-to-peer infrastructure, including the storage requirements at each participant and the amount of messaging necessary to maintain the list as participants join and leave the system. However, in most cases, each participant can maintain only partial information about the other participants. If each participant knows about the identity of a few other peers, it can always reach the other peers in the system by using the overlay and multicast schemes outlined in Sections 2.2 and 2.3.

When maintaining a list of partial members, each peer maintains the identity and information about a few of the other participants in the system, for example, maintaining the identity of about 10 of the other participants in the system. When a new peer comes to the existing participant, the new peer knows the identity of 10 of the other participants. It can query the participants discovered so far to obtain the list of the 10 other participants that they are maintaining and thus progressively discover a large number of the participating peers. The new peer can select any 10 of the new participants to maintain in its local information.

Although the number of peers leaving and joining the system would cause the information at each participant to become out-of-date after some time, each participant could learn that a peer has left the system when it try to communicate with it for any reason. In that case, the participant can query some of the remaining peers in the list to determine a fresh list of new participants. Even with a moderately small list of partial member list (e.g., with about 10 or 20 names), the system should be able to handle a highly dynamic rate of departures and arrivals.

2.1.4.3 Member Propagation with a Hint Server. We discussed a way to discover members in a peer-to-peer system by maintaining a directory of the existing members above. The directory contains the set of current members and gives information that is as accurate as possible in the presence of dynamic information. The task of maintaining accurate information in the directory service can cause quite a bit of complexity and overhead within a computer system, because every member leaving or joining the system must register with the directory service.

However, if the information contained in the directory were not necessarily consistent, but only an approximate information, the overhead involved with maintaining it would not be so large. As an example, if members only registered with the directory server when they joined, and the directory server automatically deleted their information after a day, a participant would not need to contact the directory server more than once a day, no matter how many times it left and joined the system within that day. Of course, the information in the directory server will not be quite correct but may still be good enough for the purposes of communication in an overlay network.

The directory server that maintains some information not guaranteed to be accurate is generally called a hint server. A participant could contact the hint server to find out about the identity of some (say 10) members that are currently participating in the list. Chances are that some of these members are no longer in the system (say 6 of them are not present). However, four of the existing members are still around and can be contacted. If each of those 4 members contains information about 10 other members, there is a list of 40 other members of whom some (let us say about 16) are still around in the system. By repeating this process, a new participant can discover the identity of most of the other participants within the system.

Note that all of the techniques mentioned above can be combined to produce a hybrid solution that addresses the limitations associated with each single one. Thus member propagation can be used with a DNS server-based mechanism to locate a hint server. Otherwise, a central directory service can be used, but member propagation is used to ensure that the directory service is not asked to handle queries of a large amount of information.

Once some or all of the existing peers have been selected, the new participant can proceed to create an overlay network connecting it to the other peers.

2.2 OVERLAY FORMATION

If the discovery process results in each member having the network address of all other participants, the participants can communicate with each other over the network link. However, for most large applications of peer-to-peer networks, each participant only knows a small subset of all the participants. The participant would connect to some members of this subset and thus be able to communicate directly to them. Messages exchanged between participants that are not directly connected are forwarded by the other participants until they reach the desired recipient. A logical interconnection is created among the different participants for this forwarding. This logical interconnection provides for a smaller network of the peers that is overlaid onto the global Internet. This overlay network provides the basic communication mechanism for peer-to-peer networks.

The overlay links can be created between any two machines that are able to communicate over a network. The manner in which the interconnection occurs depends on the firewalls, address translators, and other mechanisms that exist between two peers involved in the creation of an overlay. This section describes the various ways in which an interconnection can be formed between two peers and how the interconnections are used to create an overall mesh between all of the peers.

2.2.1 Creating an Overlay Link

The simplest way for a peer to connect to another peer is by establishing a direct connection between the two. If there are no firewalls, address translators, or other interfering devices between the two peers, such a direct connection is easy to establish. One of the peers needs to discover the other, discovery essentially consisting of finding the address of the machine on which the other peer is running and the port number on which the other peer is waiting to receive messages. A port is like a mailbox on the local computer that an application can check for messages, and a fixed number of such ports (65,536 of them) are available for any application to use on a machine. Once the address and port number are known, a message can be sent to the peer and connection can be established directly. Firewalls and other security devices may prevent direct access between the two peers.

2.2.1.1 Communicating Across Firewalls. Firewalls and other types of security devices are present at a large number of locations in the Internet, and they prevent unfettered access to applications that they are protecting. A firewall is usually placed to protect a private network (e.g., an enterprise network, a home computer network, or even a single personal computer) from unfettered access by the outsiders. The most common types of firewalls allow communication only to selected applications on selected machines on the private network.

For two peers to communicate with each other, one of them must wait for the other to contact it. The waiting peer must listen for incoming messages on a port number that is known to the other peer. The other peer must send a message to the waiting peer to initiate the communication. This message, which establishes the connection between the two peers, contains the port number of the sender so that the waiting peer can reply back. Most common applications tend to use standard port numbers [1] to listen for connection establishment, although one could also use non-standard port number. The port numbers of the sending and receiving applications are included in a standard header format on messages that are exchanged between the applications.

A private network connecting to the Internet would configure its firewall so that it will only allow connection establishment messages to a few selected port numbers to be received inside the firewall. If the waiting peer were listening on a port that was permitted to go through the firewall, this would be no different than being on the open Internet. Any other peer could connect to the waiting peer behind the firewall.

Most firewalls allow outward connections and will permit access from the private network to the open Internet, at least for a selected set of protocols and ports. A firewall would typically notice that a client inside the firewall has sent a connection establishment message to an external application and then allow communication between the internal and external addresses. Thus a peer behind a firewall can connect to a waiting peer outside the firewall.

Some applications use a fixed port number to listen for the initial connection establishment message but subsequently use a different port number to exchange data. If the port number that is used for the data exchange is blocked by the firewall, such applications will not be able to operate outside of the firewall. To work around this situation, (i) one could open a block of port numbers that are used for the data exchange and only use that set of port numbers for the data exchange or (ii) open up the ports on the firewall dynamically as the port numbers used for data exchange are selected. The latter option would require an automated mechanism to track the data exchanges as they are established, and to modify the configuration of the firewall accordingly. Note, however, that playing with firewall configuration in this manner would make the private network more vulnerable to a hacker on the Internet.

2.2.1.2 Communicating Across Two Firewalls. Sometimes the peers that wish to establish a connection with each other are on the opposite side of two firewalls. If none of the firewalls are configured to allow one of the peers to take on the role of the waiting/listening peer, the two will not be able to communicate with each other. In such cases, communication is usually not possible without the help of a third party that can be accessed by both of the peers. However, the presence of a willing third party server could enable communication between the two peers.

Consider the case of two peers both of which have decided to listen on a fixed port number (say port 5000) for their peer-to-peer application. They are both running on machines behind firewalls that cannot be configured to allow in connections on port 5000. This could be because the administrative control of the firewalls may be in different hands than that of the machines on which the peers are running. Let us further assume that the firewalls have been configured so that they do not allow outbound connections to port 5000 from any client inside the firewall. Both of the firewalls will only permit outbound connections using the HTTP protocol to pass through, so that clients inside the firewall could access web servers on the Internet but no other protocols or ports are allowed on outbound access.

This may sound like a situation in which the two peers would not be able to communicate at all. However, there are ways in which the two peers can communicate with each other, albeit with the help of an intermediary node that is outside both the firewalls. The intermediary must listen for incoming requests on a port number using a protocol that is allowed to pass through for outbound access from both of the firewalls. In the above example, both firewalls allow the HTTP protocol to pass through. Thus the intermediary has to run a web server on the Internet that both the clients could access. The peers would then exchange their messages by using the intermediary.

To exchange messages, each of the peers accesses the intermediary web server. A specific URL is invoked by a peer to send messages to the other peers. Each peer can also periodically access the intermediary (using the HTTP protocol) to see whether it has a message waiting for it. The intermediary web server provides scripts that perform the task of handling messages between the peers. As an illustrative example, the intermediary web server provides two cgi-bin scripts, one accessed via the URL http://intermediary/send and the other accessed via the URL http://intermediary/receive. Each of the URLs can pass parameters using the POST conventions of cgi-bin scripts. The parameters passed with the send URL identify the sending peer, the receiving peer, and the contents of the message. The send script copies the message to a queue from which the receiving peer can obtain the message when it invokes the receive URL. The parameters passed with the receive URL identify the receiving peer, and the receive script simply encapsulates the messages in the receiver's queue into an HTML document (which is the form of messages the firewall would expect to be exchanged on the URLs) and sends them as a response to the receiver.

For two peers to communicate, they must exchange messages by using some conventions they both agree on. These mutually agreed conventions define the communication protocol between the two peers. HTTP is one such communication protocol that is used by browsers to access web servers. Although a general peer-to-peer application would not use the HTTP for communication among peers, it can use the HTTP protocol as an underlying mechanism to facilitate its own communication protocol. In general, application-layer protocols are implemented by using one of the two standard transport protocols (TCP or UDP) widely available on the Internet. HTTP and the peer-to-peer protocol would both be application-layer protocols implemented normally in that fashion. However, in the presence of firewalls, the peer-to-peer application-layer protocol uses another application-layer protocol (the HTTP protocol in the above case), as if it were just a transport protocol. In other words, the peer-to-peer protocol is tunneling through the HTTP protocol in order to bypass the restrictions imposed by the firewall.

Because most firewalls would allow the HTTP protocol to pass through them, it is a good choice for tunneling other protocols. Implementing software that will allow message exchanges using HTTP as a tunnel is relatively easy, and public domain implementations of similar tunnels are readily available. The GNU httptunnel protocol [9] is one such public domain implementation.

There are two drawbacks of tunneling, one related to performance and the other related to security. Tunneling through HTTP protocols is much more expensive and cumbersome than sending a simple packet on the Internet and thus not very efficient. Also, tunneling can expose a machine to malicious attacks from outside the firewall. Because the firewall is not aware of the protocol being exchanged through the HTTP tunnel, it may not be able to perform some filtering and checking functions. If one were tunneling an E-mail exchange program through the HTTP tunnel, the firewall might not apply the traditional virus checking that it would apply to E-mail messages. Thus the machine running the HTTP tunnel could not count on the safeguards that the firewalls may be providing and would need to implement the necessary safeguards on its own.

2.3 TOPOLOGY SELECTION

In peer-to-peer applications, overlay links are used to create a mesh interconnecting all of the peers that are participating in the overlay. The topology in which the overlays interconnect the peers would have a significant impact on the efficiency of the communication among the two protocols.

Two key properties determine the effectiveness of the overlay mesh that is formed in peer-to-peer networks. The first property is the diameter of the mesh, which is the largest distance between any two peers in the mesh. The distance may be measured either in hops (the number of overlay links between two peers) or in the communication latency (time taken for a message to propagate between two peers). The other metric is the average degree of the mesh, which represents the number of direct overlay links that a peer has. Peers with a larger degree need to process more messages and incur a higher overhead in their operation. At the same time, a larger degree provides for better fault tolerance in case one of the peers were to fail. In general, one would like to minimize the diameter of the overlay mesh in order to reduce the communication latency between a pair of peers, while keeping the average degree of the peers at a moderate level.

The goal in peer-to-peer systems is to form an overlay that provides for efficient communication. One would like to avoid a situation in which all peers end up connected to each other in a linear fashion, a topology that maximizes communication latency and is susceptible to failures. There are multiple ways used within peer-to-peer systems to obtain a mesh with good operation characteristics as described below.

2.3.1 Random Mesh Formation

Formation of a random mesh interconnecting the peers is a commonly used technique that works very well with a large number of participants. During the formation of a random mesh, each of the peers randomly selects a few of the existing peers (which it has identified through some discovery mechanism) and connects to them. This results in a randomly connected graph with each node having approximately the same average degree. The drawback of random connectivity is that some of the peers randomly selected may be quite far away, for example, a node in North America may select a peer in Asia-Pacific and vice versa rather than a peer close by.

For the peers to connect to a more reasonable set of peers, many peer-to-peer implementations also use simple delay measurements between peers to select their peers. The discovery process would result in providing each new peer with a large potential number of sites they could choose to connect to. The peer would choose a subset of the potential sites (e.g., 10), connect with all of the peers in the selected subset to measure the time it takes for a single message to be sent and received from each of these peers, and then choose a few (e.g., 2) with the smallest such time. This provides for a better assurance that the overlay links created in the peer-to-peer system are between two nodes that are close to each other.

2.3.2 Tiered Formation

The random mesh formation is a good procedure to use when all the nodes participating in the peer-to-peer system are roughly equivalent, that is, they have the same type of connectivity to the network and have roughly equivalent processing power. In many circumstances, however, the nodes that participate in a peer-to-peer system may be quite different in their connectivity and processing power. A computer connected to a peer-to-peer system via a dial-up modem cannot be very efficient in the task of forwarding messages between other peers because of bandwidth limitations on the dial-up line. A computer connected to the Internet via a cable modem, DSL, or T1 line would be in a much better position to forward messages between peers.

The idea behind the tiered formation [10] is to divide computers into multiple categories or tiers and to have a overlay mesh that is structured in multiple tiers, each tier connecting to machines within the same tier or to the next tier. The simplest form of the tiered system divides the nodes into two types, the peers that are suitable for forwarding messages from other peers and the peers that are not. Machines with low-bandwidth connectivity or poor processing capabilities are not deemed suitable for forwarding messages. They can only join the overlay mesh as leaves and are allowed to join only one of the other machines that act as message forwarders. Some implementations would allow the leaf nodes to join more than one message forwarder simultaneously. The message forwarders form a random mesh among themselves, as well as connecting to the leaf nodes. The discovery process would only consist of discovering the identity of the message forwarders in the system, because they are the only ones to whom a new peer can join.

The model of two tiers can be extended to that of multiple tiers.

Nodes are divided into tiers, with tier 0 consisting of nodes that form the core of the peer-to-peer network. A formula based on machine processing power, network connectivity, and available disk storage can be used by a peer to determine the tier to which it belongs. The discovery process indicates the tier of a peer along with other information such its IP address and the port number it is listening on. The node would select to join randomly with some set of peers that have the same tier and to one (or a few) peers that are in the immediately lower tier. Such a tiered approach can lead to a more efficient overlay than random joins.

2.3.3 Ordered Lattices

An ordered lattice provides an alternative mechanism to interconnect nodes in a peer-to-peer network and has been proposed within systems such as CHORD [2] and CAN [3]. In the ordered lattice approach, nodes are arranged so that they work together to cover a space along multiple dimensions.

The easiest lattice to explain is the two-dimensional lattice. The participating nodes are arranged in rows and columns so as to form a rectangular grid. Each node maintains direct connections to four neighboring peers (the ones to the left and right of it and the ones on the top and bottom of it). In some implementations, the leftmost peer in each row considers the rightmost peer in the same row as its left neighbor, thereby forming a circular row structure; a similar circular column structure is formed by defining the topmost peer in each column to be the neighbor below the bottommost peer in the same column. Other implementations leave the peers in the edge positions with less than four neighbors.

Messages in this infrastructure are routed parallel to the axes of the lattice. A new node joining in the system must find a location for itself within the lattice. The location could be on the edge or in the middle of the lattice. The node then inserts itself into the lattice by connecting to the neighbors along each dimension. Any existing connection among the neighbors is broken, and the new node becomes the intermediary between them. It should be noted that insertions and deletions of nodes in the lattice imply that different rows or columns in the lattice have different numbers of members between themselves.

The system described above could be easily extended for more than two dimensions. One of the main applications of the lattice-based structures has been to define peer-to-peer infrastructure for storing/retrieving items by keys. Each key is comprised of a number of different parameters, each parameter mapping to a dimension in the lattice structure. Peers can navigate along the dimensions of the lattice in order to determine locate entries by their keys.

The lattice structure is more ordered than the random mesh structure and can provide a more efficient way for exchange of messages. However, maintaining the regular structure of the lattice requires more complexity and bookkeeping than the random mesh structure. The mechanisms needed to restructure the network when peers join and leave are also more complex.

In Chapter 3, we look at the use of the overlays formed in peer-to-peer networks to implement application-level multicast.

3

APPLICATION-LAYER MULTICAST

All computing machines that make up a distributed systems need to communicate with each other in order to perform their function. The most common type of communication among machines is unicast communication, in which two machines communicate with each other by sending messages addressed to the other. Other types of communications can prove very useful in other contexts. One such communication mechanism is multicast communication. In multicast communication, a machine sends a message that is intended to be received by several other machines. For systems that involve multiple machines interacting with each other (e.g., a peer-to-peer system), such communication can be much more efficient than point-to-point communication.

A special case of multicast communication is broadcast communication, in which a message is sent to all the machines in the system. In the context of peer-to-peer communications, broadcast communication would mean that a message sent by a peer is intended to be received by all of the other peers whereas multicast communication would mean that a message is sent to a subset of the peers.

Unicast communication is a basic service provided by all networks. Multicast communication can be provided at the network layer or built on top of the unicast communication provided by the

Legitimate Applications of Peer-to-Peer Networks, by Dinesh C. Verma

ISBN 0-471-46369-8

network by the peers in a peer-to-peer system. The latter is termed application-layer multicast.

In this chapter, we look at the mechanisms that are needed to support application-layer multicast in a peer-to-peer system. In Section 3.1, we look at the general approaches toward multicasting. This is followed by a discussion on the network-layer multicast support that is available within the Internet and the issues associated with using network-layer multicast. Finally, we look at the issues that arise in application-layer multicast and some systems that implement application-layer multicast.

3.1 GENERAL MULTICAST TECHNIQUES

To support multicast communication, any distributed system must provide the following functions:

- *Group Addressing Mechanism:* A name or address must be provided as an alias for the members of a multicast group. If a group consists of a small number of members, it may be possible to enumerate all of the members instead of providing for a explicit name. However, to deal with groups of scalable size, a naming scheme must be provided for multicast communication.
- *Group Maintenance:* Multicast groups are dynamic, with members joining and leaving them over time. A protocol that allows the members to leave and join the multicast group must be available. Furthermore, the current state of the multicast group (list of members and any required routing information) must be available to those nodes in the network that need it for purposes such as forwarding messages.
- *Message Forwarding Scheme:* A multicast communication scheme must provide for a way by which a sender can send messages to the members of a multicast group. This requires the provision of a scheme by which message forwarding is done. Each message should be received once by the members of the multicast group, and message forwarding must provide a way to eliminate duplicates and avoid looping.
- *Multicast Routing Scheme:* In many cases, the forwarding of messages is obtained by maintaining multicast routing tables, which contain information on how to forward messages

by individual nodes so as to perform the message forwarding function. The maintenance of such routing information is considered a part of the multicast routing scheme.

In addition to the above functions, some multicast systems also must provide support for the following functions:

- *Secure Communications:* Messages sent to a multicast group may need to be maintained securely. The tasks involved in maintaining secure communication groups require techniques for encrypting messages and mechanisms to maintain the keys required for encryption so that only the current membership of a multicast group has the set of keys needed to decrypt the messages.
- *Reliable Communications:* A multicast communication paradigm can be built with the unreliable communication model, in which a message is sent to all of the recipients on a best-effort basis, that is, recipients are not guaranteed to receive the message. Some applications would require a different semantics, in which a message should be delivered to all of the recipients in a multicast group. Maintaining reliable multicast communication requires a protocol that ensures the latter semantics.
- *Flow and Congestion Control:* Associated with the reliability issue is the topic of flow and congestion control within the network when multicast communication is used. If one is exchanging data at high rates with multicast, the multicast stream may overwhelm the buffers at intermediate nodes forwarding the messages and the receivers. Techniques to avoid buffer overflows at receivers are termed flow control, whereas techniques to avoid buffer overflows at the intermediaries are termed congestion control. Multicast communications at high data rates must implement schemes for flow and congestion control within the network.

We look at these general issues in more detail in this section.

3.1.1 Group Addressing

In most communication protocols, the sender of a message must identify the recipient of the message. In unicast communications, such identification is done by identifying the recipient in the header of the message. In multicast communication, analogous identification of the receivers must be made.

In most networks, the addresses that can be assigned as valid receivers are limited. In these cases, a subset of the valid addresses must be identified as multicast addresses. IP multicast [11] uses such a mechanism to define multicast addresses. In other cases, a unique identifier can be used as the identity of the multicast group, with some flag or other indication in the message header identifying the receiver as a multicast group address rather than an individual machine or application.

When messages are forwarded during multicast communication, the multicast address must be translated into a set of output links on which a copy of the message is to be forwarded. Usually, this requires looking up a routing table that is indexed by the multicast address. The multicast routing protocols have the responsibility for keeping the routing table up-to-date.

In multicast communication systems that are targeted at a smaller number of members, the individual members can be explicitly named in the message header. This simplifies the task required for group maintenance. XCAST [12] is a multicast communication paradigm that uses such a scheme. Explicit enumeration of members eliminates the need to have a multicast routing table, and the traditional schemes for forwarding messages for unicast communication can be used to make replicas of the message and to forward them appropriately.

3.1.2 Group Maintenance

Group maintenance refers to the task of maintaining the information about the current set of members in a group and providing a mechanism to enable new members to join and leave the group. Group maintenance can be provided in the following ways:

1. *Centralized Approach:* In a centralized approach to maintaining group membership information, a server is used to maintain the current information about the members of a group. When new members wish to join the group, or an existing member needs to leave the group, the server is contacted to make the necessary changes. The information about the group from the server can be used to build for-

warding tables. The centralized approach also provides a good way for maintenance and distribution of security keys for group communication.

The centralized server approach is the easiest and most effective to implement. However, a single central server would not be able to scale for a large number of participants. Thus a more distributed kind of group maintenance is needed.

2. *Distributed Approach:* In a distributed approach, many servers work together to maintain the information needed about the group. In the limiting case of the distributed approach, in the spirit of peer-to-peer networks, each participant in the group runs a server that is responsible for group information maintenance. Any of the servers can receive requests to update and query information about group membership. The servers exchange information with each other to maintain knowledge about the group members. The knowledge exchange could take one of two forms:
   (a) Exchange of group membership information with other servers. Each server maintains information about all of the groups in the system. When it receives an update, it modifies its state information and relays changes in its state to other servers.
   (b) Exchange of information about group location with other servers. Each server only maintains information about the local members that have joined a group. The servers do not propagate the list of members upward to other servers; they simply let the other servers know that they have information about some members of the group. In other words, each server acts as a proxy member of the group to which some members belong, receives messages to the group, and relays them to the local members.

   The latter approach is used in the Internet Group Management Protocol (IGMP) and allows for a more scalable approach by which groups containing a large number of members can be easily supported.
3. *Directory Server Approach:* In the directory server approach, one maintains a logical directory mapping the group address to all of the nodes that are members of that group. This approach works best when there are many groups but the size of each group is relatively small. The directory server could

be a single server providing the mapping from group name to the list of members, or it could be a distributed set of cooperating servers. In a distributed implementation of the directory server, each of the servers may be responsible for a subset of the multicast groups. Each server has information that can map the requests for modifying or querying the members of a group to the corresponding instance of the directory server, which can then provide the mapping to the right set of members. Within the Internet, the DNS naming scheme can be used to map the name of a machine to one or more IP addresses of a machine. Although the users of DNS to maintain multicast membership is not very common, it can provide an effective way to maintain information about groups that are relatively unchanging and contain a few members.

3.1.3 Message Forwarding Scheme

Message forwarding refers to the process of sending messages to all of the recipients so that each recipient in the multicast group receives a copy of a message. Ideally, each message should be delivered once to each of the recipients. There are a variety of techniques that can be used for message forwarding in a general multicast system.

The first method for message forwarding is to have each sender find out the members of a multicast group and to send a copy of each message to the recipients with a unicast connection. In other words, the sender simply treats multicast communication as a set of unicast communications to each of the group members. This approach would work well if the number of nodes belonging to a multicast group is small. Clearly, the approach would have serious drawbacks if the number of recipients in a multicast group were to run into hundreds or thousands. Also, this method results in sending a lot of messages over any shared network paths among the recipients.

An alternate method would be to broadcast a message to all of the recipients. Each node would check the message to see whether there are any local members of the group and then forward to the local members. If the original broadcast from the sender does not reach all of the nodes (e.g., when the nodes are connected together in a mesh), each node also forwards the message to all of its neighbors on the network so that the message is received by everyone. For multicasting to a group of nodes that are connected by a broadcast network (e.g., a local area network such as Ethernet), this approach works quite well. For nodes that are connected in a mesh with only point-to-point connections (as is common in most peer-to-peer environments), this approach can work as long as each node is forwarding messages to the others and precautions are taken to avoid looping and duplicates.

One way to avoid loops is to put an upper limit on how many times a message will be forwarded within a network. Each message contains a counter set to the limit, and the counter is decremented every time a message is forwarded. Because every message expires after some time, no messages remain circulating forever in the network. This may still result in some nodes getting duplicate copies of a message. To reduce this, a unique identifier associated with each message (e.g., the identity of the sender and an increasing sequence number) could be used to identify duplicate messages and discard a message that is seen a second time around.

The broadcast approach works well when multicast groups are large and members of a group are present in many locations on the network. In that case, the number of messages that are sent on links where there is no member present is small. However, when a multicast group contains only a smaller subset of all the nodes in a network, this approach can create many unnecessary messages in the network.

The third approach would be to maintain a routing table at each of the nodes in the network. The routing table would maintain information about which links an incoming message on a multicast group ought to be forwarded out on. The routing table can thus enable efficient broadcast on only that part of the network where members of a multicast group are present. This approach requires that an entry be maintained for each multicast group that indicates which are the outgoing links on which a message needs to be forwarded. Routing tables are usually created so as to arrange the members of a multicast group into a tree-structured topology. A tree topology has the characteristic that there are no loops within the tree. The maintenance of the trees and routing information is done by means of routing protocols described in Section 3.4.1.

Note that all of the above approaches for multicast assume that the messages are being transmitted to a group of recipients with a best-effort paradigm.

3.1.4 Multicast Routing

Multicast routing refers to the task of maintaining the tables that are used in message forwarding to determine which of the links a message destined for a multicast address ought to be sent to. These tables are created by combining the information obtained related to group maintenance along with the information required for unicast routing.

There are two ways that a tree-based routing infrastructure can be made for a multicast group. From the perspective of any member of a multicast group, one could find an optimal tree that can be used to send messages to all of the other members, with optimality defined as minimizing the latency in sending messages to the receivers or in trying to reduce the number of links that a message has to traverse. The optimal tree for reducing latency in sending messages can be obtained by finding the shortest path routes from the sender to all of the receivers and joining all of the shortest paths into a tree. The optimal tree for minimizing the number of messages can be obtained by finding the minimum cost-directed spanning tree from the sender to all of the receivers. For multicast communication, each node can maintain routing table information about one tree per sender in the multicast group.

When the number of members is large, maintaining a tree per sender could be quite expensive. A spanning tree connecting all of the members is sufficient for the purpose of sending messages among all of the participants. Sending messages along a common spanning tree may cause the messages to take a longer path than on a tree that is constructed by taking the shortest path from the sender to all of the receivers, but it results in immense reduction in the amount of information that must be maintained. To build a single tree, one defines a specific node as the core of the tree for each multicast group. Routing protocols are used to maintain the routing information about nodes that are in the multicast group. With core-based trees, a sender that is not in the multicast group can send messages to the multicast group by sending a message to the core node, which can then relay it over to the set of recipients. This technique can also be used by a sender to start sending messages to a multicast group while it is still joining the group.

The group maintenance information is combined with any routing schemes for unicast communication to determine the right set of routing tables defined either for a shared core-based tree or for a tree per sender.

3.1.5 Secure Multicast

Sometimes, the members of a multicast group need to exchange information in a secure manner. Only the members of the multicast group should be allowed to receive the messages that are being sent to the group members. External members should not be able to understand the messages even though they may be able to intercept and receive a copy of the messages. In other words, the contents of the information exchanged on the multicast group should be encrypted in some manner, and the key needed to decrypt the messages should be available only to the members of the group.

The encryption of messages is a straightforward application as long as a shared key is maintained among the group members. For a multicast group with a fixed number of members, a secret key can be established and maintained externally and used for communication. For a dynamic multicast group, where members may leave or join the group, the keys must be changed every time the group membership changes. When a new member joins, a set of authentication mechanism must be used to validate the credentials of the new member. Typically, a new member can join the group by authenticating itself with any of the existing members (or a specifically designated authentication server) and can obtain the secret key that is being used currently on the successful completion of the authentication problem.

When an existing member leaves the group, the situation can become more complicated. If the same secret key continues to be used, the ex-member can still decrypt the messages. One of the assumptions in secure multicast communication is that the members keep the keys securely. An ex-member may not be under any such obligation. This implies that the secret key must be changed whenever an existing group member leaves. The maintenance of changing keys is the critical aspect of managing secure multicast communication.

A common approach to relaying key change information is to have a group controller that is responsible for managing keys for multicast communication. The group controller will be contacted when any member leaves or joins the group. On the departure of an existing member, the group controller changes the key to be used for communication and notifies all members of the change in the current key. Usually the notification message is sent with the previous key for encryption, and all members are required to reauthenticate themselves with the group controller on the receipt of the notification message.

With multiple senders in a group, there is a time window in which the current members of a group may not know about the new keys that the controller has issued out. To address these cases, a sequence or version number is maintained by the group controller on the different keys that are issued on dynamic changes in group membership. The version of the key used to encrypt a message is included in clear with the message. When a group member sees a version that it has not seen before, it can reauthenticate itself with the group controller and obtain the new key.

3.1.6 Reliable Multicast

Multicast communication as described above operates in a best-effort manner. Messages are likely to reach all of the members of a multicast group. However, there is no guarantee that the messages will indeed be delivered at least once to every member of the group, assuming that all the members are reachable from the sender. A message may be lost or corrupted on a communication link and therefore not reach one or more of its intended recipients. Reliable multicast addresses the issue of ensuring that all the members of a multicast group that are connected to the network receive a copy of a message sent by the sender.

Three different approaches can be used to provide for reliable multicasting: sender-based retransmission, receiver-based retransmission, and reliable forwarding.

In sender-based retransmission, each recipient acknowledges the receipt of a message to the sender. The sender keeps track of the recipients that have not acknowledged the receipt of the message within a time-out period. If a large number of recipients have not acknowledged, the message is transmitted on the multicast address. If only a few recipients have missed out on the message, the sender may retransmit individually to each of them with a unicast scheme. If a fixed number of multiple retransmissions fails to reach a recipient, that recipient is presumed to be unreachable.

In receiver-based retransmission, the onus of keeping track of missed messages is put on the receiver instead of the sender. A sequence number is assigned to each message so that a recipient can identify gaps and missed messages. When a recipient notices that it has missed a message, it asks the sender to retransmit the message to it. For the last message to be sent on a multicast communication channel, the sequence number of the last message transmitted by the sender is included so that any recipient that might have missed the last message on the channel can identify the gap and ask for a retransmission. If a recipient does not receive any message after a time-out period, it queries the sender to see whether it has missed any retransmitted messages, and whether the communication has been terminated.

Both of these approaches put the onus on the sender for maintaining information about all of the receivers and/or to handle retransmission requests from all of the receivers. In reliable forwarding, the intermediaries that are forwarding messages along the multicast addressing tree take on the onus of performing the reliable communication to the other nodes that are downstream. Each link of the spanning tree is thereby made reliable, and messages are forwarded reliably along the links of the multicast tree. As long as the multicast tree is stable, the reliable forwarding by the intermediaries would result in ensuring that the message reaches all of the recipients at least once. However, if the topology of the multicast tree changes, then messages may be lost during transition as a new multicast tree is formed. In those cases, the system falls back to using one of the previous two mechanisms for ensuring reliable delivery to all of the recipients.

Practical multicast delivery protocols are built with one of the above three mechanisms or variations thereof.

3.1.7 Multicast Flow and Congestion Control

In any multicast group, all the participating nodes could have a wide variety of processing power, disk space, and available memory. In any type of communications, it is important to ensure that a powerful sender does not overwhelm a receiver by sending too many packets at once. Flow control is the process by which a sender regulates its transmission so as not to overwhelm the buffers available at the receivers. Sometimes the bottleneck may not be the receiver but a node between the sender and the receiver. Controlling the flow of packets so that no intermediary is overwhelmed is referred to as congestion control.

Flow control and congestion control are complex subjects even for the simplified case of unicast communication, and more so in the context of multicast communication. In unicast communication flow control is done by having the sender and receiver each exchange information about the number of memory buffers they have outstanding for the communication. The sender always ensures that it does not have more data in flight (i.e., data that have been transmitted but whose receipt has not been acknowledged by the receiver) than the buffers that a receiver has. In multicast communication, one could follow the same approach by taking the minimum buffer sizes available from all the receivers that are members of a multicast group. However, this forces all receivers to obtain data at a rate limited by the slowest machine, and a single slow member can reduce the throughput of the multicast communication significantly.

A similar situation applies for congestion control, which consists of detecting whether a node on the path between the sender and the receiver is congested and, if so, reducing the amount of packets being sent. Determining whether a node is congested can be done by noting when messages get lost or by having routers explicitly convey the congestion information to the senders. Because the paths from a sender to the receivers in multicast communication can return different levels of congestion information, the sender must determine how to incorporate that into its transmissions. It can transmit at the rate allowed by the slowest path, but, like flow control, this can result in slowing down receivers that have a relatively uncongested path from the sender.

One possible way to deal with the issue is to divide the multicast group members into smaller groups, so that members in each group have comparable flow and congestion control properties. Then the sender could follow the rate that is permitted by the slowest receiver. This does require the creation of more multicast groups, but the increase in the number of multicast groups may be compensated for by the fact that each group consists of receivers with comparable characteristics and so no receiver is unduly slowed down or overwhelmed by the sender. This approach is quite suitable for multicast routing schemes that maintain a tree per sender for a multicast group.

An alternative approach is to have intermediary nodes take over the flow and congestion control issues related to their immediate neighbors in the multicast tree. Each node uses unicast flow and congestion control to ensure that the path to its neighbor in the multicast tree is not overwhelmed. The intermediaries are responsible for buffering messages and relaying them to the neighbors at a rate that they can sustain. As long as the disparity in rates is not so large so as to overwhelm the buffers at the intermediary, this approach can obtain satisfactory performance.

Another approach for intermediary nodes is possible in some types of multicast applications such as live audio and video transmissions over a multicast infrastructure. For those applications, an intermediary node can reduce the amount of bandwidth to a downstream router and transmit a reduced-fidelity version of the broadcast stream. In this manner, a receiver with lower processing power would receive audio/video streams with a lower resolution transmitted using reduced bandwidth.

A combination of the above techniques is used to support multicast communication at the network layer, as well as at the application layer.

3.2 NETWORK-LAYER MULTICAST—IP MULTICAST

In IP multicast, a special group of IP addresses are reserved to denote multicast addresses, that is, addresses that contain more than one machine. There are a set of well-known addresses that define all hosts on an IP subnet or all routers in an IP subnet. Other multicast addresses can be created and maintained by means of the Internet Group Management Protocol (IGMP).

End-host computers join the multicast groups by using IGMP. When an end-host wants to join a multicast group, it sends a message to that effect to the router. All multicast-capable routers also periodically query all hosts on the subnet connected to each router interface to determine whether there are hosts that are members of a specific multicast group on it. Routers maintain information about local group members and exchange routing information with other multicast routers to create multicast packet forwarding trees in the Internet. The common routing protocols used for multicast routing are Distance Vector Multicast Routing Protocol (DVMRP) and Protocol-Independent Multicast (PIM).

In DVMRP, when a router receives a multicast IP packet, it sends the packet out of its interfaces except the one that leads back to the packet's sender. Routers that receive packets for a multicast group that have no local members or any downstream interface can send a prune request to the upstream router asking that it no longer be sent packets belonging to that multicast address. Such prune messages are valid for a specific time period, after which the upstream router can send packets over once again. Thus DVMRP follows the approach of broadcasting packets by default, letting the receivers filter out if they want to obtain a specific packet. It is designed for large multicast communication groups that span almost all of the network.

In contract, PIM is specifically designed to account for two different design points, one that is targeted for large multicast communication groups and another that handles multicast communication groups that contain a small number of members. The former is supported by PIM dense mode, which maintains one broadcast tree per sender for each multicast group. A broadcast and prune mechanism similar to DVMRP is used to maintain the trees. PIM sparse mode is the other variant of PIM, which maintains a core-based tree mechanism for forwarding packets.

To transfer routing information about the multicast group, many of the standard unicast routing protocols have been extended to support information about multicast routing. Such extensions have been made to intradomain routing protocols such as OSPF [13] as well as interdomain routing protocols such as BGP [14].

3.2.1 Problems with IP-Layer Multicast

Although IP-layer multicast offers many benefits from the point of multicast communication and is implemented by almost all router vendors, it is still not deployed widely in the Internet. Several enterprise operators are also reluctant to turn on IP multicast functions within their enterprise networks. In this section, we take a look at some of the issues that are inhibiting the wide deployment of IP multicast.

A global multicast backbone (M-BONE) exists on the Internet, but access to M-BONE is provided on an ISP by ISP basis and it is hard for an application deployer or individual user to influence that decision. For multicast to work, the ISPs need to enable multicast operation within their network. However, for operation on the global Internet, multicast must be enabled across all of the ISPs that are connecting the different computers that are required to run a distributed application. There are no good ways to discover for a set of machines whether they have multicast support enabled between them. As a result, applications that require IP multicast and need to run across multiple administrative domains would need to implement multicast capabilities at a higher layer.

Multicast communication at the network layer also has several issues related to reliable delivery. Reliable multicast transport protocols must deal with many complex issues, and there are no established standard mechanisms for their support. Although there is a working group within the IETF [15] working on such standard reliable transport protocols at the time of the publication of the book, it will be a while before those standards are established and widely available.

Multicast communication is also susceptible to many security vulnerabilities. The IGMP protocol is designed to be open, and access control for members joining the groups is not very well developed. The multicast routing protocols are also vulnerable to routing attacks, and multicast packets can easily be tampered with. The emergence of secure multicast protocols and IP security alleviates some of these concerns, but multicast security remains a concern.

Another issue that several enterprises have with enablement of multicast is the performance hit that it creates on the enterprise routers. Most commercial routers are highly optimized for unicast communication and use a variety of techniques for efficient packet forwarding. A common technique is to forward unicast IP packets by simply manipulating some header fields within the adapters of a router, whereas more complex packets are processed by the main processor in a router. Multicast forwarding is more complex than unicast forwarding and may require making multiple copies of packets in many instances. As a result, the processing of multicast packets is often relegated to a slower unoptimized path that involves the main processor in the router. High-bandwidth multicast applications, such as video broadcasts, can significantly reduce the forwarding capacity of the router. This makes many enterprise operators cautious about turning on IP-level multicast as a general service in their networks.

These limitations of network layer multicast make its implementation at the application level highly desirable. A more detailed discussion of the issues facing the deployment of multicast in networks can be found in [16]. In Section 3.3, we take a look at some ways by which multicast support can be enabled with peer-to-peer systems.

3.3 APPLICATION-LAYER MULTICAST

In this section, we will take a look at how multicast communication among a group of machines can be implemented as an application-level service. Specifically, we will look at the issue of supporting application-level multicast with a peer-to-peer infrastructure.

When a peer-to-peer infrastructure is used to support application-level multicast, we are interested in enabling many-to-many communication among a set of peers that are connected together in an overlay topology with each other. The peers have formed a mesh of overlay links connecting each other and are able to relay messages among themselves by using this mesh. This infrastructure can then be used by the participating peers to broadcast messages from one peer to all of the others.

3.3.1 Broadcast Mechanisms in Peer-to-Peer Networks

Message broadcast among peers can be readily achieved by means of flooding of messages over the overlay. A broadcast message transmitted by one of the peers can be relayed by the other peers on the other links of the overlay mesh until it reaches all of the participants. Flooding schemes require a way to avoid packets looping forever, for example, limiting the total number of hops a query will be propagated to or having each node only process queries it has seen for the first time. The former can be done by using a time-to-live field along with the query, which is incremented at every node; the query is discarded if the field increases a fixed threshold. IP forwarding uses a similar technique except that packets start with a fixed value of the time-to-live field, the value is decremented at every hop, and the packet is discarded if the field reaches zero.

If each peer must process a message only once, all peers must remember the messages they have encountered or a message must contain the list of peers it has already traversed. Both of these schemes will require space, either in the message or at a peer. In all practical implementations, a bound will need to be placed on the upper limit of this space in the message or in the node. Because a finite space opens the possibility of infinite looping, a time-to-live scheme must be used to counter a loop through the peer overlay that is larger than the permitted space. As an example, suppose that all messages contain a list of K last peers they have encountered. Any loop among peers that involves less than K overlay links will be eliminated because a peer receiving the packet twice can check for its identity in the last K peers and discard the message. However, if the loop has more than K overlay links before it reaches the peer, the peer will not be able to identify the packet as a duplicate and the packet can be propagated infinitely within the network. A time-to-live counter of K (or more) would eliminate the risk for such a looping.

When a peer-to-peer overlay mesh structure creates a tree topology to interconnect the peers, broadcast can be readily obtained without looping by sending messages on the spanning tree connecting all of the peers. A time-to-live mechanism is still helpful in these cases because some transitional loops may form when the structure of the tree changes.

3.3.2 Multicast in Peer-to-Peer Overlays

Restricted communication among a set of peers where only a subset of the peers want to communicate as part of a multicast group can be obtained in the following different ways:

- The peers that wish to form a multicast group can create their own individual peer-to-peer overlay and then use broadcast on that new overlay to communicate with each other. Peers that wish to leave the group do so by leaving the overlay, and new peers can be added to the overlay by the overlay maintenance mechanisms described in Chapter 2.
- The peers that wish to form a multicast group can use a broadcast and filter approach to receive messages. The underlying overlay is used to broadcast messages to all of the

peers. Peers that are in the multicast group receive and process the messages. Peers that are not in the multicast group simply forward the messages for broadcasting or discard them.

- The peers that wish to form a multicast group can implement multicast routing by building a full-fledged multicast function within the peer-to-peer overlay. This implies that the peer-to-peer subsystem implements one of the many different options enumerated in Section 3.1 for multicast group formation, message forwarding, and routing. Security and reliability mechanisms can also be added as appropriate from the various alternative means discussed in Section 3.1.

Because peer-to-peer groups are formed from a subset of nodes in the entire graph, it is quite common to find implementations of application-level multicast that are supported as broadcast among all the members of an application group connected together. As an example, the Gnutella protocol uses flooding for broadcasting of messages. Many other application-level multicast systems are available as research prototypes. These include YOID [17], ALMI [18], End System Multicast [19], and Overcast [20]. Most of these implementations can be readily recognized as peer-to-peer systems by the manner in which they form an overlay network and implement broadcast/multicast functions among them.

Multicast support provides a common function that can be used to build specific applications. One such application is the file-sharing application in peer-to-peer networks, which is discussed in Chapter 4.

4

FILE-SHARING APPLICATIONS

One of the most common applications of peer-to-peer networks is the sharing of files among different peers. In a file-sharing application, all of the peers in a system collaborate to share a common set of files that are collectively maintained by all of the peers. This chapter describes how file-sharing applications work. Although file sharing is a legitimate application of a peer-to-peer network, it is often used to share audio/video content in a manner that violates the copyrights of the content owner. This chapter also discusses the manner in which enterprises and ISPs can prevent the illegal sharing of copyrighted content by using peer-to-peer systems on their infrastructure, as well as the countermeasures that many peer-to-peer systems implement to work around those safeguards.

4.1 FILE-SHARING OVERVIEW

In this section, we take a look at how file-sharing applications work. We assume that each of the peers implements some variation to create an overlay network connecting all the peers. A typical file-sharing application would include three additional components needed at each peer, a disk management component, a file

Legitimate Applications of Peer-to-Peer Networks, by Dinesh C. Verma
ISBN 0-471-46369-8

indexing component, and a file search/retrieval component. The disk management component is responsible for determining which portion of a peer's local disk storage will be shared with other peers. The file indexing component determines how files are indexed and searched for within the files located locally or on another peer. The file search/retrieval component provides a mechanism by which a peer can search for files that are not present on the local disk. This mechanism involves a way to broadcast queries to the other peers and to receive their responses back on files that match the queries.

In addition to the above three components, a file-sharing application may also include components for ensuring access control, components for digital rights management, and some enhancements for making the file search process more efficient, for example caching or parallel downloads from many different machines. These components are described in the next subsections.

4.1.1 Disk Space Management

A peer may choose to share all of the local hard disk with other peers, or it may choose to share only a fraction of the local disk. In most cases, it is better not to share the portion of the local disk that contains the operating system and other critical files. The shared file system can be written by other peers in the file-sharing application, and exposing critical system files can lead to a serious security vulnerability.

The disk management component of most file-sharing systems is relatively simple. A user is simply asked for the file system that is to be shared with other peers on the network. A default file system that is shared is included in most file-sharing systems and would often default to a known directory.

The shared file system could see changes from two possible sources. Whenever a user at the local peer places a file into the shared file system, it would become accessible to the users on the remote peers. A part of the shared file system, however, would be automatically managed by the file-sharing application. The file-sharing application would keep local copies of any files being searched by the user on the local peer or keep a cached copies of other files in order to provide them to remote users. Such files would usually be kept in a default directory of the shared file system. A index file that catalogs all of the files that are being shared with remote users may also be created and kept automatically by the file-sharing application.

4.1.2 File Indexing

On a single computer, we typically tend to look for files by their names. However, in a distributed file-sharing environment, different peers on different machines may assign different names to the same file. As an example, a song titled “Any Song” by a singer “John Doe” may be stored in one machine with the name of john-doe.mp3, while on another machine it may be stored with the name of anysong.mp3.

One can build a distributed file-sharing program that shares files only by using their names. In this case, the file-sharing program allows users to store and look for files with the same name anywhere among all the peer machines. However, given the fact that the same file may be named differently by different people, this type of file-sharing program would be very restrictive. To look for a file, one would need to know its name or some fraction of its name. In many cases, a user may wish to look for a file with a specific type of content as opposed a specific file name. As an example, a user in a file-sharing program would like to find all presentation foils available in Microsoft PowerPoint that deal with the subject of peer-to-peer networks.

One way to provide for limited content-based searching for files is to define several sets of attributes for each file. Some of the attributes of a file are often known to a computer, for example, the date a file was created or last modified. Some other attributes of a file can be determined by using some heuristics, for example, the suffix of a file can be used to determine the format in which files are contained, or the title of a web page can be inferred by looking for text contained between specific tags in the HTML representation. Other types of attributes can be specified by manual annotation, for example, many music recording programs allow their user to annotate the title and artist of a song. Automated attribute determination software to determine the title and keywords for many types of file formats (e.g., Adobe Acrobat files, Microsoft Word files, Postscript files) can also be used to obtain attributes for files created in their respective formats. Such software programs use a variety of heuristics to guess at the attributes of the files from their contents.

When attributes are associated with the file, the file-sharing program can store the attributes as metadata associated with each of the files in the shared system. The attribute metadata can be stored in a special index file in each directory of the shared file system. When files are copied across peers in the file-sharing applications, their attribute information is also copied over.

4.1.3 File Search/Retrieval

When a user of a file-sharing application wants to look for files with a specific set of attributes, he or she provides the type of attributes he/she is looking for via a user interface at one of the peers. The peer looks through the local shared file system to see whether there are any files that match the specified set of attributes. If so, the local files are returned. If no local files matching the attributes are found, the search continues on to the other peers on the system.

The search on other peers is usually done by broadcasting the query to all of the other peers. The broadcast can be done by means of flooding or by forwarding requests along a spanning tree. Some peer-to-peer systems may build spanning trees based on the attributes that are specified for a set of files. Each of these queries is processed by the receiving peers. If any of the files match the criteria specified in the query, the peer sends a response back to the requesting peer. The response may be sent back directly, or it may be sent back by using the reverse path on the same set of overlay links that were used by the query to propagate from the requesting peer to the responding peer. A good optimization in the system is for a peer that has a file matching the desired attributes not to propagate the query further.

On receiving the response from all of the peers, the requesting peer can select one of the responding peers and download the best matching file for the response from that peer. If there are multiple responses, the requesting peer may download all of the matching files in the respond, download one that is deemed to be the best match by some heuristic or manual selection, or download a few of the responses. The requesting peer would typically copy the files in its local shared file system and thus would be able to provide this file to any other peer that may be looking for the same file in the future.

4.1.4 Access Control and Security

Thus far, we have only considered the sharing of files among peers without any need for authentication or access control on the files. However, it is relatively straightforward to augment most peer-to-peer file-sharing systems for access control on selected file systems. Access control can be provided either by having each peer maintain its own local access control guidelines or by acquiring access control guidelines from some trusted authorization agency.

When maintaining access control lists locally, each peer can independently determine which of the other peers gets access to a file and which peers are not allowed to get access. A peer can authenticate the identity of another peer before giving it the contents of a file. Each peer receiving the file would have to agree not to forward the restricted file to any other peer or would obtain the set of authorized peers for the copied file and only allow the authorized peers to obtain the replica of the file. The identity of a peer can be authenticated by the peer presenting credentials that are certified by a third party that all the peers trust. When the access control list for a file changes, a peer would only accept those changes that come from the peer that gave it the original copy of the file.

A simpler way to provide for access control functions in peer-to-peer computing is to have a trusted third party that provides for access control and security in the traditional manner. To get files from another peer, each participating peer must present a certificate that is provided by the trusted third party. The peer that is serving out the file validates with the third party that the requestor is entitled to receive a copy of the file and provides the file only after the security checks succeed.

4.1.5 Anonymous File Retrieval

In some cases, participants in a peer-to-peer system want to obtain files anonymously, that is, the receiver of a file does not want to reveal its identity to the peer that serves out the file. When anonymity is desired, the peers need to hide their IP addresses from other peers. One way to do so will be to have files transferred not via a direct connection between the sender and the receiver, but along an indirect path. In most peer-to-peer networks, the broadcast query reaches a peer via several intermediaries. Although each intermediary knows the IP address and thus the identity of each peer with which it has a direct connection, it does not know the identity of the other peers.

One way to obtain anonymous communication is described in SRIRAM [21]. Each peer assigns a random address to peers to which it is directly connected. The mapping of the address to the real neighbor in the link is only known to the local node. The SRIRAM architecture creates a spanning tree to provide a broadcast capability among all the peers.

Anonymous communications are always done by using the spanning tree path instead of directly among the involved parties. Two chains of link-local addresses are included in anonymous communication, each chain encoding an anonymous path from one sender to the other. The only exception is anonymous queries on the spanning tree, which are used to discover the initial chain of link-local addresses to use.

A participant looking for available resources anonymously will send out a query to be broadcast among all the neighbors. Each participant will include the link-local address of the neighbor from which it has received the query before forwarding it on to the other branches of the spanning tree. The link-local addresses are appended to a growing chain of the path to the requester and form the anonymous path back to the querying participant. When a participant wants to send a response to the query, it removes the last link-local address from the chain and sends the message to the neighbor with the specified link-local address. These responses are also subject to the reverse path accumulation and create a reverse path to the respondent.

For the anonymous communication to work, we only need at least one of the intermediary nodes to play by the rules and not reveal its mapping of link-local addresses to others. Each node is aware only of the identity of its immediate neighbors and is not able to infer the identity of any other participant unless all of the members of the spanning tree along its path collude with it. As the number of participants increases, the ability of any individual to obtain such colluding members becomes negligible.

An alternative option to maintaining local name bindings and reverse path accumulation is to have all the peers maintain a connection open to their peers for outstanding queries. The information to get the response back up one hop is known at each peer, and responses are communicated hop by hop on open connections at each of the peers. This eliminates the complexity of reverse path accumulation but imposes an extra burden on each of the peers.

Both of the above approaches can be seen as adaptations of the general notion of onion routing [22] for anonymity.

4.1.6 Search Acceleration Techniques

When the size of a peer-to-peer system becomes large, the search for files may take a long time. Several enhancements and modifications have been proposed so that the file searches on peer-to-peer networks can be accelerated and improved upon. In this section, we look at some of the proposed techniques.

Because file searches are focused on broadcasting the query to as many nodes as possible, the topology of the underlying overlay plays an important part in how quickly the query propagates forward. The creation of a more efficient overlay can improve the time it takes for a query to propagate in the system. Some of the proposed techniques classify all of the peers into one or more categories [10], each category containing peers with comparable processing power, network connectivity, and memory. Others assign ranks based on similar attributes in each of the peers [21] The overlay is formed with the more powerful set of nodes in the center, which makes the broadcast process faster. The commercial system Kazaa [5] also divides peers into two categories, one only acting as leaves and the other acting as message forwarders. The choice of whether a peer acts as a leaf or as a message forwarder is determined by the user of the system, but the guidelines request users with good network connectivity and powerful systems to act as message forwarders.

Another common technique for improving performance is the use of caching at different nodes in the peer-to-peer system. An unofficial 80-20 rule seems to exist in most areas of computer science, which states that most (80%) issues arise only from a small number (20%) of the involved entities. In the context of peer-to-peer file-sharing applications, the 80-20 rule indicates that most of the searches in a system are looking for only a small fraction of files in the system. If the popular files are replicated widely, then overall search time for files in the system decreases. In most peer-to-peer file-sharing systems, the client that receives a copy of the file also allows it to be served out to other peers. In other words, each receiving peer is caching a copy of the file. A copy of the file can also be cached at other peers in the system. If the file was transferred by multiple peers (e.g., as in the case for anonymous communication), then each of the intermediary peers can cache the file for future queries.

Even when a file is sent directly to the requesting peer, it is possible for other peers to cache information about the peer that had files matching a specific query. A peer-to-peer implementation may have the peers with files that match a specific query propagate that information on the path that the query traversed. Peers cache the contents of the query and the identities of the peers that have files matching those contents. On subsequent queries, these results can be used to locate the content quickly, eliminating the need for broadcasting a query further when nodes with matching content are already known.

Another technique for improving the time to download a file to a peer is downloading a file in parallel from multiple different sources. Because a single file is likely to be replicated at many peers, the contents of a file can be downloaded faster by opening simultaneous connections to all of those peers. As an example, consider a file that is 50 blocks long and is available from 5 different sources. A peer would download the first five blocks of the file in parallel from each of the sources by opening five concurrent connections. The source that is quickest to finish is asked to serve out block 6, and subsequent blocks are requested in the same manner. As a result, more blocks tend to be downloaded from the closer sources, and the effective time to download a file is also reduced significantly.

4.1.7 Digital Rights Management

The rights of the author or creator over information created in any medium are recognized by most of the member countries of the United Nations. When an electronic document is created, the creator of the document may want to maintain some rights over the document as it is disseminated. These rights include the ability to prevent copying, alteration, or unauthorized transmission of a copy of the document to other entities. Depending on the type of information, the authors may want to assert a subset of the rights they have. Creators of religious text, for example, want their information to be freely copied, as long as the copy does not modify the original content. Creators of some software programs allow unlimited copying and use for noncommercial purposes but want to receive some payment for commercial usage of their software. Some creators of information may want to exercise all of their rights, preventing any type of copying or alternation of their information except from a few authorized sources.

Digital rights management systems allow the creator of a file to specify which of the operations are permitted to other users of the file and what types of operation are prohibited. These rights can be associated with files as metadata or embedded within the content of a file. When a file is obtained by a peer, a digital rights management component in the receiver can check whether any digital rights are associated with the document and enforce those rights. Digital rights might prohibit caching of files in some instances. Some digital rights management systems include the rights information in the content of the file itself and require the use of special viewers that enforce the rights to use that file.

Peer-to-peer systems can implement a digital rights management component, which can prevent copying or servicing of files from unauthorized sources. The lack of widely acceptable standards for digital rights specification is an obstacle toward that goal. Some of the efforts in that area includes the Open Digital Rights Language (ODRL) [23] and extensible rights markup language (XrML) [24]. Other related work includes the Secure Digital Music Initiative (SDMI) specifications [25] and interoperability of data in eCommerce systems (INDECS) specifications [26].

4.2 USAGE OF FILE-SHARING APPLICATIONS

File-sharing applications can provide a very effective system for storing and sharing files across a group of users. File-sharing applications can be used in the context of a limited group of users, as well as in public peer-to-peer forums.

A group of researchers or students working on a project can use a file-sharing application to share the results of information that they have found. Traditionally, some collaborative work has required the creation of a common repository where such information is contained. Such collaborative work spaces could be created as a common web server location, a Lotus Notes database, or even a common directory on an accessible server. However, the maintenance and updating of the common repository is a chore that often gets overlooked by the participants. A peer-to-peer file-sharing application can obtain the same goal and requires much less manual administration. All files that individual users create on specific file systems on their machine becomes accessible to the other users. The drawback of the peer-to-peer system would be that the files are only accessible when a specific peer is available and connected, although copying and caching of files to other peers could reduce that drawback to some degree.

File sharing also enables access to a vast amount of information that is available among all of the computers in the participating users without requiring any central coordination or management. Therefore, all employees within an enterprise may be able to share technical notes, documentation, and associated information with each other without the need to maintain a centralized repository of the same information. Thus information exchange is enabled without requiring significant overhead for managing a server or site that stores the information.

In a more public forum, peer-to-peer systems have been used to widely share information of mutual interest to a large group of users, including sharing of music, videos, books, and other type of information among users of systems such as Kazaa [5], Morpheus [27], and Gnutella [28]. When information is shared with the consent of the creators of the information, such systems provide a wonderful way of looking for hard-to-find items.

4.2.1 Limitations of File-Sharing Applications

Despite their widespread use and many useful characteristics, file-sharing applications suffer from some limitations and drawbacks. The key characteristic of the peer-to-peer file-sharing application is that it does not require centralized management. However, the same characteristic makes some things difficult and limits the applicability of file-sharing systems. Some of these limitations are described below.

- *Versions and Latest Information:* Although file sharing applications work well for content that is finished and unchanging, they do not usually do a good job of managing different versions of a file and replacing old versions with newer versions of the content. Thus older versions of a file may linger around in the system even when more current information is available. If the information were available at a single authoritative server, managing the versions would be a much simpler task.

- *Spurious Data:* In a peer-to-peer system, it is relatively easy for pranksters or malicious users to insert spurious or incorrect information in different files and to annotate them with attributes that are misleading. Thus a search for information in a peer-to-peer system may find several spurious files before it can obtain information that is authentic. One way around this problem would be for creators of content to sign the files digitally, by computing a hash of the contents so that the originator of a file can be validated.
- *Legitimacy:* When file-sharing applications are used to share content that violates the copyrights of content creators, the users of the file-sharing system may be violating the copyright laws of the local jurisdiction. Even when file-sharing applications are enabled for sharing legitimate files, there are no easy ways to prevent placement of copyrighted content among the set of files being shared. Guidelines on what type of content is to be placed on the system could work well for a private group of users, for example, a set of enterprise users, but are difficult to enforce in an open group.
- *Security:* Peer-to-peer applications download files from unknown peers and users. If such downloaded files contain viruses and Trojan horses, and a downloading peer software is not adequately protected by virus-checking software that can identify and isolate the virus, the peer may be infected with the virus, with consequential damage to the system and loss of productivity. The security of a system can also be compromised by users inadvertently sharing private files that they should not have been sharing.

Despite these limitations, file-sharing applications can be used to advantage in many different contexts, as long as precautions are taken to work against the limitations of the distributed nature of the system.

4.3 PREVENTING UNAUTHORIZED FILE SHARING

Unfortunately, the predominant use of file-sharing applications at the time of writing this book involved the exchange of copyrighted music files among users. This has led to a series of legal actions by producers of music records to prevent the spread of peer-to-peer systems. The music producers are involved in a series of litigations against developers and users of peer-to-peer software systems. The developers of many peer-to-peer systems, on the other hand, feel that they have an obligation to help free exchange of information among the users. Without taking any position on the relative merits of the arguments on either side, one can still assert that many enterprises and networking service providers run the risk of potentially costly litigation if peer- to-peer systems are used for illegal file sharing on their networks.

To guard against such litigation, most enterprises and service operators have put policies in place that prevent the users from deploying file-sharing applications. To enforce these policies, effective policing mechanisms must be in place within the system that monitor file-sharing software. At the same time, the creators of the peer-to-peer systems have taken countermeasures that work against some of those safeguards. In this section, we take a look at some of those measures and countermeasures that can prevent the use of illegal file-sharing applications.

Avoidance of a lawsuit is not the only reason for trying to control peer-to-peer file-sharing applications. Because of their immense popularity, file-sharing applications have become the major fraction of traffic on the Internet. Many universities are finding that a disproportionate share of their bandwidth is being used by peer-to-peer file-sharing applications, which can interfere with the performance of other applications such as web servers and electronic mail transfer. In these cases, it makes sense to attempt to regulate the amount of bandwidth a file-sharing application uses on the network.

Another reason to prevent file-sharing applications in a public forum are the security threats that can arise from the use of a open file-sharing application. Because anyone can place files on an open file-sharing system, malicious users may attempt to put viruses and Trojan horses in the files that are downloaded by other peers. These viruses can damage a computer or read and transfer private files outside the firewall. Improper use of file-sharing applications can also expose sensitive private files to unauthorized users.

The measures that can be taken against undesired file-sharing applications depend on the perspective and control of a network operator. In this section, we look at the countermeasures in three contexts, that of an enterprise, that of a network services provider, and that of an external user.

For the enterprise context, we look at the operator of an enterprise network that has a policy against enterprise users participating in a peer-to-peer network. The enterprise would typically be connected to the external Internet by means of a firewall. The goal of the network operator in the enterprise is to put in safeguards to prevent users from running a peer-to-peer service that connects to the external Internet as well as to prevent users from running a peer-to-peer service that may be used only by users internally within the enterprise.

The network service provider is an organization that provides network connectivity to its customers. The policies and guidelines of the network provider may preclude the operation of peer-to-peer services on the customer's premises. The network service provider is only able to view the aggregate traffic coming out of a customer's network or computer and thus has a coarser view of the applications that are running on the customer's premises. The customer may be an enterprise with its own internal network, a home user with a single computer connected to the ISP, or a small business- or homeowner with a small local area network.

An external user does not have administrative control over any part of the network but is able to participate in peer-to-peer file-sharing systems. The external user can then attempt to disrupt the operation of the file-sharing systems by using a variety of techniques. In essence, the external user is a malicious peer in a file-sharing application that is trying to disrupt its operations.

4.3.1 Firewall-Based Techniques

The operator of an enterprise network has administrative control over the firewalls that connect enterprise users to the external Internet. The firewall provides a point where the enterprise operator can disrupt the operations of peer-to-peer networks, or at least those file-sharing systems that interact with external users. More firewalls are capable of blocking off traffic to a set of external computers by using their IP addresses, blocking traffic that uses a specific set of port numbers, or a combination of both.

If a file-sharing application requires access to a well-known server for creating its overlay, or to search for files present on peers, it is relatively easy to block access to that system. The service of Napster, which was subsequently shut down after court orders to that effect, involved using well-known servers to access the location information about different files. Most network operators at enterprises and universities found it easy to block Napster by blocking access to the IP addresses used by Napster. The newer versions of peer-to-peer systems do not use a centralized server and thus cannot be blocked with that level of ease.

If a file-sharing application uses a set of well-known ports for communication, it is relatively easy to block access to those port numbers from any computer inside the network, as well as to prevent access to those port numbers from any computer from inside the network. As an example, the file-sharing applications of Kazaa and Morpheus use the TCP port number of 1214 by default, WinMX uses port 6257 for TCP communications and 6557 for UDP communications, etc. Blocking access to those port numbers between external and internal computers is likely to prevent those applications from running. The solution is effective but not perfect. The administrator has to obtain a list of ports for all the file-sharing applications, and there are a large number of these file-sharing applications to contend with. File-sharing applications may change the default port numbers or may run with other port numbers. Furthermore, some legitimate applications that use dynamic port numbers may be disrupted inadvertently. As an example, file transfer usually creates a data transfer connection using any open port number on the client. If the port number selected happens to be the one blocked by the firewall because it was used commonly by another file-sharing application, some legitimate file transfers may also be blocked.

One way to prevent legitimate applications from being blocked is to keep the port open but assign upper limits on the amount of bandwidth that can be transferred in or out of the port. Several firewalls and traffic control devices are available in the market that can monitor specific connections for the amount of bandwidth used and restrict usage to be below a specified bandwidth threshold. Assigning an upper rate on the bandwidth of the ports would permit legitimate applications to run at a reduced rate and also restrict peer-to-peer applications from using up too much bandwidth. This scheme can be quite useful when the goal of the network operator is to reduce the amount of bandwidth used by peer-to-peer applications, rather than to terminate peer-to-peer applications completely.

To work around the restrictions that may be placed by firewall operators, some file-sharing applications have started to use a dynamic set of ports to establish their connection. The discovery process for these applications often consists of a list of IP addresses and port numbers that can be used for a new client to connect to the existing peers. A new peer tries the listed machine and port number combinations in turn until it gets a successful connection to an existing peer. Once connected, the set of currently active peers can be updated by a client. Because file-sharing peers are using a random set of port numbers to communicate with each other, a firewall cannot easily block then on the basis of port number alone. Legitimate clients and applications can use those same port numbers, so blocking a large set of port numbers is not advisable for firewalls. However, if the firewall is able to identify both the IP address of the machine and the port number, it can block communication to and from that machine with the specific set of port numbers to thwart the file-sharing application. The use of dynamic ports in peer-to-peer computers is often termed port hopping.

To thwart port hopping, the firewall administrator must learn the current set of port numbers and machines that are involved in a given peer-to-peer system. It can learn about them in several ways, the simplest way being to run a copy of the peer-to-peer application on a machine owned by the firewall administrator. The machine can be configured to make the shared file space nonexistent (e.g., map it to /dev/null on a Linux system or on a very small disk partition on a Windows machine) so that it is effectively prevented from any type of file sharing. However, this peer still participates and discovers other active peers in the system. This information can be passed to the firewall dynamically, which can block access to the selected set of machines on the specified port numbers.

The only drawback to this scheme is that the firewall must be capable of handling the large number of machine-port number combinations that will be obtained because of this participation. Furthermore, dynamic manipulation of a firewall's filtering list via a computer program is something that must be done with caution.

If an ISP maintains a firewall between its network and the access point of the customer, the ISP can use the same firewall techniques to identify and prevent the operation of unauthorized file-sharing applications.

4.3.2 Asset Inventory

Although firewall-based techniques are a good way to prevent internal machines from communicating with external machines in a file-sharing application, firewalls cannot detect and prevent file-sharing applications that happen within the enterprise. Firewalls are typically not present on the data path between two machines in the same enterprise. Even if an enterprise were to deploy several internal firewalls, they would only be able to detect file-sharing applications that run across the firewall, and not those that are all running on one side of the firewall.

One approach that an enterprise can use in most instances is to take an inventory of all of the software programs that are used on computers that run on its network. Most computers within an enterprise can be required to install a software asset inventory program that periodically checks for all the software packages that are installed on that system.

Software asset inventory programs use a variety of techniques to determine whether a software is present on a system or not. On most operating systems, each installed program would modify some system entries to register its presence. On the Windows 2000 operating system, for example, most applications would create entries in the software registry with a standard set of fields to define their configuration information. Other software programs come with software packages that have a standard name or configuration file. Even if the program does not create any entries in the system registry, or its configuration information is stored in a different location, the names and hash codes for standard install packages can be used to detect their presence. Because peer-to-peer file sharing applications would typically tend to be started automatically on a system, an examination of the standard places where programs are scheduled to run automatically can reveal their presence and configuration information. The registry entries, typical filenames and their hash codes, and other characteristic information associated with the applications are collectively called its signature.

Most of the inventory agents contain a relatively extensive list of common applications and the typical signatures that correspond to those applications. The inventory agent would go through the set of signatures and discover which of the signatures are found on the machine. The set of discovered applications is then reported back to a main server. If a peer-to-peer software package is detected on any of the machines, the enterprise administrator can request the owner of the machine to disable that software. Users that fail to comply may find their network access being blocked by the enterprise firewall, or other disciplinary measures can be exercised.

The asset management approach can only be used by enterprise customers. It cannot be used by networking services providers, who may not have the ability to place inventory software on a customer's machines.

4.3.3 Port Scanning

In some environments, it is not possible to install inventory applications on the end-computers. This would be quite typical in environments like the university, where students and faculty typically own and administer their own computers, unlike the corporate environment, where computers are often owned and managed by the corporate IT staff and standard software configuration can be enforced on the clients.

Port-scanning software provides a way to detect the presence of unwanted applications on machines in such an environment. A port scanner is a program that tries to connect to all of the ports on a computer system to see whether a server application is active on any of the ports. Once a server active on a port is identified, the scanner program tries to figure out which application it really is. It performs this determination by trying out the initial contact messages of the known applications and checking whether the server responds with the right answer. Such a scanning process will discover all active instances of an application. Such probing techniques are used in security products that analyze network vulnerabilities, and they should work to identify the operation of peer-to-peer networks. The only caveat is that the scanner should know the protocol used by the application. For peer-to-peer implementations that use a well-known protocol such as Gnutella, this scheme would work very well. For peer-to-peer applications that use proprietary protocols, the scanner would only be able to report that an unknown application is operational on the network.

Port scanning can most efficiently be used by enterprise operators. An ISP can also use the port-scanning software but may not be aware of all the different computers that are present on a customer's network. Scanners also can generate a significant amount of load on the customer network, which can cause performance problems for the customers. Therefore, port scans are not the best tool for a networking services provider to detect file-sharing applications.

4.3.4 Usage-Based Rate Control

Usage-based rate control techniques are a way to provide peer-to-peer applications from using up too much of networking resources. This prevents the peer-to-peer applications from impacting the performance of other applications running on the network.

The basic idea in usage-based rate control is to estimate the average amount of bandwidth that a user would need when it is not running any file-sharing applications. An enterprise operator can compute the total amount of bandwidth that a typical machine may need in a reasonably large time period, for example, on a daily basis. Typically, such usage would be a relatively small fraction of the maximum bandwidth that the machine can sustain. The network operator can then assign a daily quota on the maximum amount of bandwidth a machine is entitled to have. This quota could be a few multiples (e.g., 5–6 times) of what the needs of the average customer/machine would be. The network operator would monitor the usage of the network from each of the machines on a frequent basis (e.g., every hour) and obtain a list of machines that have violated their quotas. The users that have violated their quotas would have their network connectivity blocked by a firewall for the rest of the day. Alternatively, if a traffic control device is present on the network, the operator can place limits on the maximum amount of bandwidth any machine is allowed to have on the system. An ISP may similarly estimate the total amount of bandwidth that a typical customer may use in a day without running any peer-to-peer applications and monitor/regulate customers that exceed their bandwidth quota in a similar manner.

The monitoring and regulation may be done in the incoming bandwidth, the outgoing bandwidth, or a combination of both.

The approach does not prevent peer-to-peer file-sharing applications from running but reduces their bandwidth usage within manageable limits. If these limits prevent heavy usage of the peer-to-peer file-sharing applications, they may keep the enterprise or network provider below the radar screen of music producers who may bring litigation against heavy users of file-sharing applications. One advantage of this approach is that bandwidth control is independent of the specific type of file-sharing applications and does not require any customization for new file-sharing applications. The downside of the method is that it may unnecessarily penalize users that may be needing extra bandwidth for other, legitimate purposes on the network.

4.3.5 Malicious Participation

An enterprise operator, an ISP, or an external user can participate in an open peer-to-peer file-sharing system as a malicious user that is participating mainly to disrupt the operation of the peer-to-peer network. As mentioned above, most peer-to-peer systems provide for a way to discover a set of potential other peers one could connect to. Thus the system allows a malicious user to discover the location and identity of other machines within the system. The malicious participant can further try to discover the identity of other users in the system by a variety of methods. Some peer-to-peer systems allow for the discovery of the participants by letting peers query other peers about their neighbors. Other peer-to-peer systems have peers directly connect to requesting peers to upload files. In the latter type of systems, the malicious user can circulate common queries to learn the identity of the users that are participating in the system. It can also keep track of the responses from the other peers to identify the placement of files among other users.

The use of a malicious participant was one of the methods suggested above to identify the set of machines and ports whose connection ought to be blocked by an enterprise operator or a network services operator. A malicious participant could also identify internal participants in the file-sharing application to an enterprise operator, who can take steps to have owners of those machines disable the offending application. A malicious participant can also inject spurious files within the network, which reduces the quality of files that the file-sharing application is exchanging among the users. Furthermore, malicious clients could inject viruses or other malicious software in downloaded files that can try to terminate the file-sharing application on the infected machines. In some cases, malicious users create valid copies of files to be shared within the network but put watermarks or other indicators in the copy of the file. These marked copies are only available in the file-sharing applications, and the presence of the marked copy of the file on any computer would serve as evidence that the computer was engaged in a file-sharing operation. Inventory software running on the machine in an enterprise environment could subsequently identify the participants in a file-sharing application.

There are countermeasures against all of the tricks that a malicious peer can engage in. Much peer-to-peer software now comes with bundled virus-checking software that can thwart attempts to insert malicious code on a peer's site or to take any other unwanted actions. Modern peer-to-peer systems can also implement a reputation or ranking scheme of different clients. Clients that are serving out files that are valid and well-formed are given a higher rank and develop a reputation of being more trustworthy because they have delivered more bytes of trustworthy files. Peers that are serving out files that are malformed are given a lower reputation/rank and are not considered trustworthy sources for downloading files. Thus malicious users trying to inject bad files into the system are given a bad reputation, and they lose their ability to spread bad files within the system.

Anonymity schemes like the ones described above in this chapter prevent malicious users from discovering the identity of people who are serving out files in the system, except for the immediate set of neighbors. The discovery of the overall network topology can also be prevented by not allowing peers to disclose the identity of their neighbors to any nonadjacent peer.

A peer-to-peer system can even thwart digital watermarks. Such watermarks are often placed in parts of image, audio, and video files and provide for a level of granularity indistinguishable to human users. These types of information can usually be corrupted slightly without any perceivable difference to a human end user. Typically, watermarks are placed in the lower bits of graphics images (which only have a minor influence on what is visible on the screen) or higher compression coefficients are placed in audio/video data (which are not audible/visible to human users). A peer-to-peer system can replace those lower bits/higher coefficients with random data on every file that it uploads from the other peers. This destroys the watermark but has only a small adverse effect on the quality of the file. If a watermarking scheme uses a scheme utilizing higher bits of image data, or lower compression coefficients, the uploading peer can inject selected random errors in the uploaded image. The random errors would not be perceptible to the human end user but would have a high probability of breaking the watermarking system.

Nontechnical approaches have also been used by associations of music producers to thwart the growth of file-sharing applications that violate their copyrights. These include high-visibility lawsuits brought against developers and users of file-sharing applications. The main motive behind these highly publicized lawsuits has been to dissuade other people from becoming new users of unauthorized file-sharing applications.

To summarize, there is a game of cat and mouse between the groups that would try to prevent the spread of unauthorized file-sharing systems and the developers of the file-sharing systems. The nature of file-sharing applications will change over time as they address the current challenges imposed by malicious users. However, there are many other applications of peer-to-peer networks that do not violate any laws of the land. Several such applications are described in subsequent chapters of this book.

5

FILE STORAGE SERVICE

The file storage service is an application that has received a lot of academic interest in peer-to-peer networking, although its practical use is not as widespread as that of the file-sharing application described in Chapter 4. The file storage service allows the storage of a file at one or more locations in a peer-to-peer network by using a uniquely defined handle for each file. The handle is also used to retrieve the file from one of the peers. As opposed to file sharing, where users are looking for files containing a specific content, the file storage application provides for searches only on the basis of the handle. This allows for a much more efficient implementation of searching for a file either for the purpose of updating or for the purpose of retrieving the file.

The file storage application has many different usage scenarios, and each usage scenario has its own specific needs that must be addressed over and beyond the basic file storage functions. One of the uses of the file storage system is to build a distributed file system from peer-to-peer applications, which provides for a much more enhanced file system than one from the local user. Another common use of a peer-to-peer file storage service is to enable anonymous and decentralized publication. On the Internet, almost anyone can publish the content he or she wishes to on a dedicated website. However, if one were to publish a sensitive document, that is, a book that may go against the established tenets of

ISBN 0-471-46369-8

a tyrannical government, anonymity would be highly desirable and a peer-to-peer file storage system would provide a better option.

The general file storage application associates each file with a unique handle and maps each handle to a unique peer in a distributed system. Any peer in the system can locate the file by using the handle, and peers with the right authentication privileges can update the file by using the handle. The file storage system must provide for two basic components, the first one associated with creation and management of the handles for the various files and the second consisting of the use of the handle to retrieve and store a file. Each of these two components is described in the following sections, followed by some usage scenarios of the file storage application.

5.1 HANDLE MANAGEMENT

Handle management refers to the problem of creating and managing file handles for files in a distributed storage system. When a new file needs to be stored into the distributed peer-to-peer system, a unique handle for the file must be created. Before a peer can retrieve the file, it must find out what the handle associated with the file is. The common techniques for creating and distributing handles in file storage systems are discussed in this section.

The file handle of a file must be unique among all of the files that are present in the system. To create a unique file handle among many different peers in a computer system, the following techniques can be used:

- Assign a unique identity to each peer and then use a unique local name in combination to provide for a unique handle. This scheme has often been used for distributed systems when the number of machines is relatively static and unchanging. A common scheme is to use the global fully qualified domain name of the machine along with the local file name or a locally generated nonrepeating sequence number. However, in a peer-to-peer system made up of client machines or home computers, machines may not have a unique fully qualified domain name. When most clients obtain an IP address with protocols like DHCP, the network address of a machine is not guaranteed to be unique, and this approach may fail to come up with a unique handle in many cases.

- Compute a unique hash of the contents of the file. It is common to use cryptographic algorithms to compute hashes over a stream of data. These algorithms generate a small hash of the contents of a file that have a very small probability of occurring once again. The SHA-1 hash algorithm [29], as an example, computes a 160-bit digest for any file that contains less than 264 bits. This limit can be considered close to infinite when you consider the fact that a file with a size of 100 gigabytes contains only 233 bits and is only about one thousand trillionth of the limiting number. For a file that is not going to be changed subsequently, a unique hash of the contents will provide for a unique identifier. This unique identifier also has the nice property that it can be used to validate the contents of a file that is retrieved from a remote peer.
- Compute a random unique identity for each independent user within the network of a fixed large size (e.g. 1024 bits), and then associate a random number to each file as a prefix to the identity of the machine. This can be considered a variation of the first approach, where the probability of two machines having the same random identifier is extremely small.

One of the key properties of a handle obtained by the last two methods outlined above is that the handle has a fixed length. The fixed-length handle allows efficient methods for locating where the file is created.

When a user wants to locate a file, he/she must obtain the handle of the file so that it can be used to retrieve the file. The creator of the file may pass the handle around to the receivers via an external mechanism, for example, by means of an E-mail message, or by means of an external site. One of the key assumptions in file storage systems is that the key is available to any user interested in locating the file.

One can envision situations in which a user does not have the handle but wants to search for a file by its contents. There are two possible approaches to searching for a file by content:

The first approach is to broadcast the message to all of the peers in the network. If the underlying network overlay supports a broadcast mechanism, a broadcast can be sent to all of the peers, and a peer with the appropriate file can send the response back to the requesting node. This would be identical to the way files are searched for by content in the file-sharing application.

The second approach is to have some mechanism by which clients can obtain the file handles of files that contain their specific content. A well-known index site could return the set of handles that correspond to specific type of contents, in the same manner that websites like Google or Yahoo return a set of URLs for a specific type of content. Alternatively, a set of index files can be maintained with well-known file handles. The index files contain an enumeration of the contents of different files by their handles. A peer looking for content can then find the set of file handles for specific types of content.

Most file storage systems proposed in the literature assume that the handle is available to the reader as well as the publisher of a file, assuming that handle information flows from the publisher to the readers via some external communication mechanism.

5.2 RETRIEVING FILES WITH HANDLES

The more complex part of a file storage application deals with the technique used for locating files based on handles. Each file must be placed onto some node in the distributed system, and the handle of the file determines the peer on which it is received. Each node may be assigned the ownership of one or more sets of handles. When a file is created in the system, it must be directed to the node that will have the ownership of that handle. Similarly, when a node needs to locate a file by handle, it must find the node that has the ownership of a specific handle.

One way to locate nodes by handle would be to do a broadcast search on the overlay making up the peer-to-peer system. However, broadcast searches are expensive, and having handles can allow for a much more efficient way to locate the node with the message. Accordingly, we will look mostly at the techniques that have been developed to route requests from any arbitrary peer in the system to the peer that contains a specific handle.

In all of the methods described in this section, a node that wants to create a new file in the storage system creates a “file-update” message that contains the handle and the contents of the file and sends the “file-update” message in the system according to the routing scheme described below. The message propagates through the different peers until it reaches the peer that has ownership of the specific handle. The peer then creates a new file with that handle locally or updates the existing copy of the local file. A "file-search" message similarly propagates through the system to the right owning peer, that then sends a copy of the file back to the sender. The copy may be sent back directly to the requesting node, or it may retrace the path taken by the "file-search" message, depending on the design of the peer-to-peer system.

In effect, all the peers in the system work together to provide for a distributed hash table, in which the handle of a file (a numeric value of a fixed size) is used to store and retrieve the file. In the following discussion of various ways to implement distributed hash tables, we further assume that each peer has an identifier with the same size as that of the handle. An identifier for a peer can be generated randomly or by computing a hash (e.g., a SHA-1 hash) over some attribute that uniquely identifies that peer.

5.2.1 Circular Ring Routing

The simplest way to route requests with the handle would be to arrange all of the peers in a circular ring so that the peer with the highest identifier considers the peer with the lowest identifier as its successor. Peers are given ownership of a set of handles. A common scheme is to assign each handle to the peer that has the longest prefix match with the handle and to select the peer with the highest identifier in case of multiple such peers. Alternatively, one could assign each handle to the node that has the smallest identifier whose numeric value is greater than that of the handle. Each node in the ring maintains information about the identifiers of its predecessor and its successor.

To route a request to the right node, the originating node first checks whether it has the ownership of the handle. If so, the request is processed locally. Otherwise, the originating node compares the absolute difference between the handle and the identifiers of the predecessor and the successor. The request is forwarded to the node that has the lower absolute difference of the two.

As an example, let us consider four nodes and identifiers that are 4 bits long. The identifiers can thus range between 0 and 15. Assume that the nodes have identifiers of 0, 4, 8, and 12. For this example, let us further assume that node 0 would have ownerships of handles 13–15 and 0, node 4 would have ownership of handles 1–4, node 8 would have ownership of handles 5–8, and node 12 would have ownership of handles 9–12. Suppose a request for handle 10 is received at node 4. Because it does not belong to node 4, it would check for the better of its two neighbors 0 and 8. The absolute difference between the identifier at node 0 and the handle is 13, and the absolute difference between the identifier at node 8 and the handle is 5. The message is then forwarded to node 8. Node 8 does not own the message but repeats the check to forward the message to node 12, which owns the handle.

Circular rings work because the absolute difference between the handle and the node will keep on decreasing as the message propagates along the ring. The approach is simple and will work for a small set of peers. It is easy to insert new peers into the circular ring. However, the circular ring approach suffers from a significant practical limitation: The messages can take a long time to go around the ring if the number of peers is large. Because real-life peer-to-peer systems can have thousands of members, the circular ring is not a practical approach.

An alternative approach to the circular ring is to have each peer maintain a table that describes the set of handles that are owned by each of the peers. To locate a specific handle, a peer simply looks up that table to find the right peer. This approach requires only one hop to find any handle, but it requires a large routing table space at each node. Thus it trades off the routing information maintained at each peer for the number of hops it takes to route a message.

Most of the practical approaches provide different levels of trade-off between the routing information required at each peer and the number of hops needed to route the message to its final destination. Many of these approaches use the circular ring as the underlying technology in combination with a smaller routing table to obtain this trade-off.

5.2.2 Plaxton Scheme

This algorithm [30] provides for a reduced number of hops by maintaining a subset of the routing information needed for all of the nodes. If there are N peers, they are all assigned an identifier of 1 to N randomly. A handle is assigned to the peer that has the longest matching prefix between the handle and the peer's identifier. Each peer maintains a routing table that determines where the search for a handle ought to be routed next.

The Plaxton scheme reduces the size of the routing table required by dividing the identifier of each peer into multiple levels. If we convert the identifier of the peer into some base (e.g., an octal number or a hexadecimal number), each level corresponds to a digit of the identifier in that base. Let us say that there are k digits in such a representation of the identifier when represented in base b. Each node maintains routing information about some b – 1 nodes that have an identifier with the first digit different than its own, another set of b nodes that have the same first digit in their identifiers as its own, but with b different second digits, another set of b nodes that have the same two first digits in their identifiers as the node's identifier, but with b different third digits, and so on. For any given identifier, each node has routing information about a node that matches at least one digit of the identifier more than its own identifier. To locate a specific handle, a peer looks up the routing information in the table with the longest matching prefix. Each node forwards the handle lookup request to a node that matches at least one more digit in the prefix than it does, and so the request can reach the peer that owns the handle in at most k hops.

As an illustrative example, let us assume that the peers are assigned numbers in the decimal base, and there are less than a thousand peers in all. Each peer has an identifier that is 3 digits long. In this example, a node with an identifier of 346 will need to maintain routing information about 8 other nodes with identifiers of 0 XX, 1 XX, 2XX, 4XX, 5XX, 6XX, 7XX, 8XX, 9XX, where X is any arbitrary digit, 8 other nodes with identifier 30X, 31X, 32X, 33X, 35X, 36X, 37X, 38X, 39X, and 8 other nodes with identifier 340, 341, 342, 343, 345, 347, 348, and 349. For the sake of illustration, let us assume that X is arbitrarily set to 0 for the above node. To route a message looking for a handle of 257, the node 346 would send it node with identifier 200. The node 200 would have equivalent routing information about 24 other nodes, which for the sake of illustration would be 000, 100, 300, 400, 500, 600, 700, 800, 900, 210, 220, 230, 240, 250, 260, 270, 280, 290, 201, 202, 203, 204, 205, 206, 207, 208, and 209. This node would route it to the node with identifier 250. The node 250 would have a routing entry for 24 other nodes, and one of these would be for the node with identifier 257, to which it would route the message as its final destination.

The routing tables as described above must be updated every time a new node joins or enters the system. Automatically updating implementations of the above scheme have been provided by the CHORD [2], Pastry [31], and Oceanstore [4] projects. The typical implementation would use the Plaxton routing scheme in conjunction with the circular ring technique. Each node is assigned an identifier that could be computed as a hash. Each handle is mapped to the node that has the identifier greater than its own value. All the peers in a circular ring are sequenced in increasing order of the identifiers assigned to them, circularity being achieved by connecting the peer with the highest identifier to the peer with the lowest identifier. Each peer knows its predecessor and its successor in this ring. A new peer computes its identifier, finds the current peer that would own that identifier, and inserts itself between the current owner and its successor. It would then obtain the routing tables by combining the routing information from its predecessor and successor.

5.2.3 CAN Routing Algorithm

An alternative way to retrieve handles in a peer-to-peer network is provided by the CAN routing algorithms. This approach maps each handle to a point in a multidimensional hyperspace. Each peer is assigned ownership of part of this hyperspace and stores the handles that are available in this hyperspace. Each peer needs only to be aware of the peers that have ownership of adjacent portions of the hyperspace. The last node along each dimension of the hyperspace is connected to the first node along that dimension to create a circular ring. Thus all nodes are arranged in multiple circular rings along a number of independent dimensions.

To route a request for a handle, each peer determines the coordinates of the handle in the hyperspace and checks whether it is within its ownership. If so, the request is processed locally. Otherwise, it is forwarded to one of the adjacent nodes. The adjacent node is selected so that it makes the request come closer to the node that would have ownership along at least one dimension in the hyperspace.

Each request is thus routed along one of the many circular

rings toward the destination node until it reaches its destination. The schema may take more hops to reach the destination, but its maintenance and upkeep are much simpler than those of the Plaxton schemes described above.

5.2.4 Modified Network Routing Schemes

Because networks have traditionally used routing tables to direct requests toward specific hosts, it may appear that similar routing techniques would be able to work for peer-to-peer overlay networks. Specifically, if the handle of a file is mapped to specific peers in the system, one could conceive of creating tables at each of the peers that direct requests toward the peer that has the actual file.

As opposed to the Plaxton, circular ring, and CAN schemes, the routing approach decouples the structure of the overlay network with the specific handles that are stored at each of the nodes. Each node must maintain a set of routing tables that direct the requests coming for a specific handle toward one of the many overlay links that have been created with other peers. Simple modifications to the existing intradomain routing protocols can be used to construct the routing table dynamically.

A common intradomain routing protocol that can be extended in this manner would be RIP [32]. In the extended version of RIP, each peer will maintain a set of routing tables that maps the handles of specific files to one of the many overlay links that the peer maintains with other peers. Any time a new handle is added to or deleted from the routing table, or if the routing table is modified, the updated routing table is sent to the adjacent peers. Each routing table contains the handle of the file whose requests are being routed, the number of hops (or some other distance metric) from the current peer to the peer owning the file, and the next peer to be used for that handle. On receiving an updated file from a neighboring peer, each peer recomputes its routing tables to see whether the best route to a specific file handle has changed. If so, the peer recomputes its own routing table and propagates the information to the other neighbors.

Routing schemes like the one described above work well if the number of handles in the system is small and the overlay connecting the various nodes does not change too much. In a dynamic peer-to-peer system where nodes come and leave frequently, such a routing scheme may not work too well. Another concern would be the size of the routing tables with a large number of objects. In the Internet, routing tables are kept within limits by using hierarchy and aggregation of prefixes in IP subnets. For routing tables to be manageable, the same would need to be done for the peers in the overlay itself. One way to obtain this hierarchical relationship is for each joining node to be assigned a set of identity ranges, all beginning with the same prefix. The prefix, padded with zeros, is the identity of a parent peer of the node. A new node may connect to multiple existing peers, but only one is designated its parent. This would mimic the behavior of routing tables within the Internet.

5.2.5 Modified Broadcast

The need for a relative stable overlay and large sizes of file handle space makes the viability of network routing schemes in the context of file storage application somewhat doubtful. A broadcast search on the entire overlay would help in identifying the node that owns a specific handle, but such searches can be quite expensive. However, there are some techniques that can be used to improve the effectiveness of the search. If responses to queries traverse the same path from the responding node to the node originating a query for a specific file handle, the intermediate nodes could cache that information and respond quickly to the subsequent queries by directing the query immediately to the right node instead of repeating the broadcast. Caching improves the efficiency of the search process.

A modification to the broadcast process can enable intermediate peers to build up their routing tables based on the usage of the files. When queries and responses are received on the same path, a peer could simply record the outbound node to which queries looking for a specific file handle ought to be directed. The most active file handle information is kept in the local routing table, and requests that match an entry are forwarded to the appropriate adjacent peer in the overlay. If no match in the routing table is found, the peer simply reverts to a broadcast search.

Some peer-to-peer systems like OceanStore [4] use probabilistic routing techniques to maintain their routing tables. The system maintains a fixed set of routing entries per adjacent peer. The first routing entry contains information about all the handles that are available locally, the second routing entry for an adjacent peer contains information about all the handles that are available locally and at the adjacent peers, the third routing entry for an adjacent peer contains information about all the handles that are available at peers within two hops along the direction of that peer. The information of handles is maintained as a Bloom filter, a compact representation of a large set of handles. A Bloom filter deploys a set of hash functions and uses that to probabilistically test whether an element is present in a large set of elements. The bitmap obtained from applying the hash function to the element will very likely have specific values whenever the element belongs to the set. Each routing entry uses Bloom filters to represent the set of handles reachable via a peer at specific hops, and a search is directed toward the adjacent peer that has the handle reachable via the shortest hop. The probabilistic algorithm may not always work, and therefore such a routing scheme must be augmented by another type of routing (e.g., one based on Plaxton's scheme) that is deterministic.

Yet another variation on the broadcast scheme is to use a depth-first search for broadcasting messages on the overlay network, which results in reducing the number of messages sent to locate a file handle at the expense of increased latency. The depth-first search can be combined with caching techniques to provide a hybrid solution for routing requests.

By a conjunction of file handle management techniques, and one of the request routing techniques described above, a file storage application can be readily built atop peer-to-peer systems. In Section 5.3, we look at some typical uses of such file storage systems.

5.3 MISCELLANEOUS FUNCTIONS

To provide for a robust file storage service, any implementation must provide for two more components in addition to handle management and routing services. A file storage service may need to support access control to prevent unwanted reading or writing of the different file systems, and it must also provide for reliability in the presence of failures of individual peers. We look at these two components in this section.

5.3.1 Access Control

When a file is created by a user, only the user or a person authorized by the creator of the file ought to be able to modify the contents of the file. In other words, write operations to files should be validated to ensure that the person updating the file has the requisite permissions to do so. Similarly, the creator of a file may only want some members of the group to be authorized to view that file. Retrieval operations using a file handle must check for the appropriate permissions before handing the file over to a reader.

Proper access control can be enforced relatively easily if all the peers can be expected to be well-behaved. In that case, each handle could be associated with an access control list. The access control list would enumerate the set of users that are entitled to read or write a file. The peer where a file is located would check the credentials of the peer requesting the read/write/replace of the contents pointed to by a file handle against the access control list before allowing the operation to proceed. The credentials could consist of a certificate issued to the user by a trusted certificate authority. The certificate could consist of either the identity of the peer requesting the operation or a role that is assigned to the requesting peer. A role allows peers to request files from any other peers without revealing their identity.

If a peer is not behaving according to the rules, the above-mentioned scheme for access control would not work very well. To guard against malicious peers, the creator of a file could encrypt the contents with a secret key. The key is only provided to the set of users that are authorized to read/write the contents of the file. Unauthorized users may be able to obtain access to the file, but will not be able to read or write it. If the creator of a file wishes to allow a set of users to be able to read the file but a different set of users to be able to modify the file, an encryption scheme with a public-private key pair can be used. The creator of a file generates a private key and a public key that work together. A file is encrypted with the private key and located at the peer determined by its handle. The peers that are entitled to read the file are given the public key so that they can validate and decrypt the contents of the file. The peers that are entitled to modify the file get the private key. This scheme will prevent unauthorized reading/writing of the files stored with untrusted peers.

Note, however, that the scheme presented above does not preclude a peer from deleting files with a specific handle stored with it. A reputation-based scheme could be used as a partial safeguard against these peers. In the reputation-based scheme, different peers would rank each other on the basis of the reliability of the contents of files obtained from the other peer. A peer that is providing valid files (those that are encrypted with the correct key of the author) is provided a higher rating than the peer that is providing invalid files. A reputation metric of all the peers can be obtained by computing the fraction of valid files served by each peer compared with the total files served by that peer. Each peer computes the ratings of other peers independently and drops connections to a neighboring peer that fails to satisfy specific thresholds of reputation.

5.3.2 Availability and Reliability

The individual nodes that make up a peer-to-peer system are inherently unreliable. If the peers consist of clients' personal computers and laptops, any peer may be turned off or become disconnected at any time. Under those circumstances, the files stored at that peer would not be available to other nodes within the system. To become a practical service where a file is ensured to be available with a high likelihood, the overall system must provide mechanisms to have the files be available with a high degree of probability.

The underlying mechanism to provide a higher level of reliability is replication. Files are replicated to more than one peer so that they become available even when some peers are not connected to the system at any given time. Replication can be performed at the granularity of a file or at the granularity of different blocks within a file.

When replication is done at the granularity of a file, each file is copied in its entirety to more than one peer. As mentioned above, many file storage services would assign an identity to each peer. Instead of an identity being associated with a single peer, many peers may be assigned the same identity. The peers with the same identity act as a cluster and contain a full replica of the files stored at any of them. One method for a cluster to operate would be to have the peers elect one of the peers as the active peer. An heuristic such as using the peer with the smallest IP address (or the peer that joined first) can be used to elect the active peer in the cluster. The active peer is responsible for participating in the file storage application protocols, and other peers act as standbys. The cluster peers monitor each other (including the standby), and a new active peer is selected if the current active peer becomes disconnected.

Routing schemes based on modified broadcast do not have an association between the handles stored at the peer and the identity of the peer. In those cases, files can be replicated on a randomly selected set of peers. The modified broadcast scheme would locate one of the replicas of the files when it is requested in a read operation. A writer, however, would need to find all of the replicas and update them with the latest version. Replicas on peers that are disconnected may not get updated. Peers rejoining the system after disconnection would need to query for newer versions of the file and see whether their local copies need to be updated. Readers would also need to check for the version of different copies of the file they find and use the latest version.

An alternative mechanism to obtain a higher level of reliability is to replicate files at a granularity of blocks, a block being defined as a fixed size of storage. Each file is composed of one or more blocks. The system may store multiple copies of each block of file or use erasure codes [33] to convert the file into an encoded version. In the encoded version of a file using erasure codes, the files are converted to a larger number of blocks that are derived by performing a sequence of operations on the existing blocks of data. The properties of the transformation allow the original file to be reconstructed by retrieving a subset of the total blocks generated. If the converted file were n blocks long, and the original file m blocks long (with m less than n), the file could be reconstructed by any peer that is able to retrieve any k blocks of the encode file (with k typically between m and n).

Higher availability is obtained by replicating each block of the encoded file a given number of times. When looking for a file, all of its blocks are searched for, either by assigning a unique handle to each of the block or by doing a broadcast search. After the minimum required number of blocks is retrieved, the original file can be reconstructed. The scheme would work even if some of the blocks were not available at the time of querying the file.

5.4 USAGE SCENARIOS

In this section, we look at some of the scenarios in which distributed file storage service has been used. Each of these scenarios requires some enhancements in addition to the basic file storage service to customize for the special needs of that usage.

5.4.1 Distributed File Systems

A distributed file system provides the ability to store and retrieve files from a set of many computers. The file system appears to be a local file system, whereas the files in the file system may be stored anywhere in the set of different peers.

To build a distributed file system, we need to be able to impose a hierarchical file structure on the objects that are stored in the file system in addition to being able to manage handles for the files, provide access control and authorization support, and retrieve/update files with specific handles. Of these issues, all except for the hierarchical structure have been addressed previously in this chapter.

To impose a hierarchical structure on the file systems, the notion of directories needs to be supported. A directory is a node of the file system that contains other directories or files as children. Each directory can be represented as a text file that contains the name and other attributes of the files contained within the directory. If one of the attributes stored in the text file is the handle associated with each file name, the directory structure is easy to impose by looking up the files or directories by their associated handles. The other attributes associated with a file may include information about how many blocks it contains, information about where the blocks might be located, or information about which handle to use to retrieve the blocks when erasure codes are being used.

To avoid performance penalties, a machine would need to cache parts of the directories and files in active use as local copies. Caching, in conjunction with other techniques mentioned above, could provide for a usable peer-to-peer model for a large volume file storage service.

The peer-to-peer file-sharing system has the advantage that the file can be accessed from any machine that the user may happen to access in the peer-to-peer environment. Furthermore, with the shared disk capacity of multiple computers on board, no computer is likely to ever run out of disk space in such a distributed file system. Two trends, however, may weaken the case of building a distributed peer-to-peer file system. The first trend is the increasing disk capacity among modern computers, which makes disk space plentiful for all but a handful of users. The second trend is emergence of centralized services on the Internet which provide some function, e.g., GoToMyPC (www.gotomypc.com).

The OceanStore [4] system and the PASTRY [31] system are examples of file sharing systems that are geared toward developing a distributed file system.

5.4.2 Anonymous Publishing

One of the key features of peer-to-peer systems is the lack of any central control. This aspect allows for publication of data by anonymous authors and the ability of a peer to retrieve any documents from other peers in the system anonymously. A peer can take on three roles, namely, that of an author, a reader, or a distributor. The author creates a file and publishes it anonymously. The reader is a peer that retrieves the file from the system. The distributor stores the file locally and provides it to any reader that wants it.

In a truly anonymous system, the readers should not know who wrote an article that was retrieved or the identity of the distributor that provided the article. Similarly, the author should be unaware of the distributor(s) that are distributing that article and the readers that are retrieving that article. The distributor should also be unaware of the author's identity and the identity of any readers of the file.

Anonymity between readers and distributors can be obtained by means of anonymous routing techniques described in Chapter 4. In anonymous routing, queries are propagated along the overlay links between neighboring peers in the system. A peer is aware of the identity of the neighboring peers but does not know where the response to a query comes from, or who originated the query. This protects the identities of the readers and the distributors from each other.

Anonymity between authors and distributors can be obtained by using the file handles to store a file within the system. The file handle protects the author's identity from the distributor of the file. However, because the file handles are computed by the author of the article, the location of distributors that use the file handle to store can be inferred in some routing schemes that assign files to peers based on the file handle. Examples of these schemes are the circular ring and the Plaxton scheme and its variants. If the identity of the distributors did not need to be hidden from the author, these schemes would be adequate. If the identity of the distributor is to be hidden from the authors, a modified broadcast routing scheme would need to be used. To protect the anonymity of distributors from authors, one scheme could be for the authors to broadcast files to be stored to all the peers, with each peer deciding randomly whether it wants to store the file and act as its distributor. The file is then distributed from various peers, but the author is unaware of their identities.

The Freenet system [34] is an example of an anonymous publishing service built on a peer-to-peer infrastructure.

6

DATA BACKUP SERVICE

One of the key IT challenges in any enterprise environment is to have an efficient and reliable data management infrastructure. A typical enterprise IT environment consists of many computing elements such as workstations, personal computers, laptops, and departmental servers. Each of these computers has a large amount of data generated during its normal course of operation. The data on these computers are often critical to the operation of an enterprise, and the loss of data can result in reduced productivity, lost revenues, and an interruption in the normal flow of business processes.

The loss of data is not an unusual phenomenon, and it occurs because of various causes, such as disk crashes, accidental erasures, user mistakes, and viruses. Data management systems provide the ability for users to recover data that may be lost for any reason. The key functions of data management systems include making automated periodic copies of data in a reliable manner and restoring data from the backup copies when required.

The traditional data management approach within an enterprise has been to operate a backup server to maintain backup copies of data across the enterprise. Each backup server handles the backup functions of a set of computers (backup clients) and must have the capacity to store all of the data present at all the

Legitimate Applications of Peer-to-Peer Networks, by Dinesh C. Verma
ISBN 0-471-46369-8

clients assigned to it, as well as providing a very high degree of availability. Consequently, backup servers tend to be high-end, expensive machines with high-capacity data storage devices. Software products that provide automatic backup and restore functions are available from many vendors, some examples being Veritas Backup Exec [35] and Tivoli Storage Manager [36]. The operation and maintenance of backup servers is a large expense for most enterprises. It is estimated that enterprises spend three dollars to manage data for every dollar spent on new storage hardware [37]. A data management system that is able to reduce the cost of data management would help in reducing IT costs in enterprises significantly.

Peer-to-peer systems require no central control and thus may provide an alternative approach that would not require the operation of an expensive backup server. This can offer equivalent functionality at much lower costs. In this chapter, we look at such an application.

We begin the chapter by examining the typical structure of a centralized data backup center. Then we describe the design of a peer-to-peer data backup service, concluding with a look at the circumstances under which a peer-to-peer data backup service might be feasible.

6.1 THE TRADITIONAL DATA MANAGEMENT SYSTEM

We assume that a traditional backup system operates in an environment in which there are multiple backup clients that are backed up at a single backup site. For scalability reasons, the backup site may consist of a cluster of backup machines rather than a single machine. Because the backup site has to handle the storage needs of a large number of clients, it is usually implemented as one or more high-end servers running with a very large amount of storage space, which may be obtained by a mixture of tape drives, storage area network devices, or network-attached storage. However, the cluster essentially appears as a single server to the backup clients.

Large enterprises may have several backup sites, with each backup site supporting a subset of client machines. The partitioning of client machines among different backup sites is determined primarily by the proximity of a client to the backup site. In many cases, the backup sites act as peers to each other, providing mirroring capability to provide disaster recovery support for storage backup functions. An example of such mirroring capabilities is provided by the IBM TESS [38].

Each backup client in the system would typically have a copy of client software that enables a user to manually back up and restore files from the servers. Additionally, the client software would typically contain a scheme to run automated backups at scheduled regular intervals, for example, each client may automatically be backed up every 6 hours. To speed up the process of backing up data, backup may be done in an incremental mode in which only changes to an existing file since the last backup are sent to the server. The metadata associated with a backup, for example, the last time that a file was backed up, is usually maintained on the backup server. The backup client would check whether the file had changed since the last backup time and, if so, would send the modified files to the storage server. In the case of full backup, all the files are copied over to the new server.

The bulk of the storage costs associated with data management in an enterprise arise from the need to manage and operate the backup site, which needs to be highly scalable and available round the clock. The costs associated with establishing, maintaining, staffing, and operating the backup server make the backup storage many times more expensive than the cost of the client storage.

The high costs associated with backup of servers leads to a paradoxical situation in which the total cost of ownership of a megabyte of storage increases as the capacity of client disks increases. As the capacity of disks on individual workstations increases, the enterprise must provide for increased storage at the backup site. On one hand, users and application developers feel that they need not worry about a few extra megabytes on their PCs because of the large capacity of the disks. On the other hand, they come under increasing pressure from the enterprise IT staff to reduce the sizes of actual storage usage, mailboxes, etc., to reduce the costs at the backup site. Because backup sites tend to be upgraded at a much slower pace than the turnaround time of PCs and laptops, we are headed for an era when storage space at backup servers is going to be expensive and in short supply, while there is a surfeit of unused disk space on each individual user's machine.

6.2 THE PEER-TO-PEER DATA MANAGEMENT SYSTEM

If most of the users in an enterprise have machines that have relatively low disk utilization, they could provide backup and restore functions to their peers in a distributed manner. A peer-to-peer paradigm for data backup and restoration would eliminate the need for a backup site and would result in substantial cost savings for the enterprise. With the peer-to-peer backup paradigm, each file is copied over to another peer in the system, rather than to a central backup site. However, the peers are not likely to have the same degree of availability as the backup site. Thus, to obtain a higher availability for each file, a file would need to be copied to more than one peer in order for it to become available.

The peer-to peer-data backup architecture on any machine creates an independent area on each peer that is used for data backup from other peers. The user of the peer can specify configuration properties like the maximum fraction of the disk space to be used for backup functions and the location of the file.

In the peer-to-peer system for data management, a common software would be installed on each of the computers within an enterprise. The common software would provide the ability to back up and restore files as needed by a workstation client. The structure of the common software consists of the following components:

- *Basic Peer-to-Peer Broadcast Mechanism:* These are the basic components available as the building blocks for performing an application-level broadcast on a peer-to-peer network. The basic peer-to-peer search mechanism provides the ability to search for a file with a given name and set of attributes on a peer-to-peer network. Libraries providing this capability are generally available as components in most peer-to-peer software distributions.
- *The Peer Searcher:* The peer searcher is built atop the basic peer-to-peer search mechanism and is used to search for a peer that would be a suitable candidate for maintaining backups of a specific file in the system. When a file needs to be backed up, the peer searcher component floats a query on the peer-to-peer network, looking for possible peers that should receive a backup copy of the file. The peer searcher components on other machines respond to such queries. The

peer searcher would then select a subset of the responding peers as suitable for replication and backup.

- *The File Searcher:* The file searcher is built atop the basic peer-to-peer search mechanism and is used to search for peers that are holding a backup copy of a file and are currently available.
- *The Backup / Restore Manager:* The backup/restore manager is the component responsible for the completing the backup of files on the local machine, and restoring a file from its backup copy, when needed.
- *The Properties Manager:* The properties manager is a component that keeps on tracking the properties of a current machine to assess its suitability to act as a backup copy for a peer. The properties manager keeps track of aspects such as when the peer is typically up and connected to the enterprise network, the disk utilization on the system, and the speed of the processor on the system.
- *The data manager:* The data manager is the component that maintains copies of backup files on the local system. This component is largely similar to that in traditional backup systems but has the additional role of maintaining the files securely so that the backed-up data is only visible to the owner of the data.
- *The schedule manager:* The schedule manager is responsible for initiating the periodic backup of the files on the local machine to remote machines on a regular periodic basis. This function is unchanged from that in corresponding traditional backup clients running on an automated schedule.

Each of these components (except the basic peer-to-peer search mechanism) is described in more detail in the following sections. The basic peer-to-peer search mechanism is described in detail as a component of the file-sharing application in Chapter 4.

6.2.1 The Backup/Restore Manager

The backup/restore manager is responsible for creating the backup of a file system to one or more peers within the system, as well as for restoring the lost copies of a file in the system. In our architecture, each file is copied independently of the other files on the system. This ensures that each file is copied to an independent set of peers.

To copy a file, the backup/restore manager first searches for a set of suitable peers for backing up the file by contacting the peer search module. The peer search module selects a set of peers and returns them to the backup module. The backup agent then contacts the peer to create a copy of the file on each of the selected peers. If the other peer already has a previous copy of the file, an incremental backup of changes to the file is made to the peer. Otherwise, the full file is copied to the backup server. The file can also be encrypted as described in Section 6.3 on security considerations.

When a copy of a file is to be restored, the date of the last restoration and the name of the file are sent to the file searcher module. The file searcher module looks for the copy of the file that is the most recently available copy among all the peers. That copy is then used to create the restored version of the copy.

The backup/restore manager is also responsible for classifying files on the local system into different categories. Files classified as application files or temporary files will not be backed up at all. Other files will be backed up in the usual fashion. Such classification is a normal feature of most traditional backup systems.

6.2.2 The Peer Searcher

The peer searcher is responsible for locating a few peers for each file that can act as potentially good sites to create a backup copy of the system. The peer searcher is easier to implement on a peer-to-peer search mechanism that implements application-level multicast than on a distributed hash table paradigm.

To search for a suitable peer, the peer searcher module floats a broadcast query using the basic peer-to-peer broadcast mechanism. The broadcast query contains the name of the file being backed up, the size of the file, the name of the local machine, and the uptime cycle to which the current machine belongs. The uptime cycle of the machine is described in Section 6.2.4 on the properties manager . The peer searcher modules on the other peers receive the query and compose a response consisting of the following properties: the uptime cycle of the peer, the amount of free space on the peer, and a Boolean flag indicating whether a copy of the file already belongs in the peer being contacted. The peer searcher module on the originating node collects all the responses and assigns a weight to each of the responding peers. The weight is computed so that peers with an existing copy of the file are preferentially selected, peers with the same uptime cycles are preferentially selected, and peers with smaller disk spaces are preferentially selected. The set of peers is then used to create backup copies of the file.

6.2.3 The File Searcher

The file searcher module is responsible for finding the existing copy of a file to be restored. The name of the local peer and the file name are used as the keys to locate the file on the existing peer-to-peer network. Each peer that has a copy of the file being searched responds to the original peer, including the time when its backup copy was created in the response. The querying peer selects the backup peer whose copy is the latest one before the restoration time. The information is passed to the backup/restore manager, which actually restores the file. The file searcher can be implemented over the distributed hash table paradigm (using the identity of the node and filename as keys to the hash table) or over a broadcast/multicast paradigm for building peer-to-peer systems.

6.2.4 The Properties Manager

The properties manager module is responsible for keeping track of the characteristics of the local machine on which it is running. The properties manager keeps track of the times during the day when the local machine tends to be up. All machines maintain a uptime cycle property. The uptime cycle is a vector of 24 numbers, each being a numeric probability that the machine will be up during that time of the day. The probabilities are computed by the properties manager, keeping track of whether the machine was up or down during the specified hour over the duration of the previous month. If the statistics are not available for a month, the statistics are computed over available data if more than 7 days' worth of data is available. If sufficient data is not available, the uptime cycle is initialized to contain a probability of 1 during 9 to 5 local time and a probability of 0 during other times.

In addition to the uptime cycle, the properties manager also keeps track of the amount of disk that is allocated on the local peer for the task of backing up files from other peers. As the disk allocated approaches saturation, the peer is less likely to respond to requests from other peers to backup copies of files from them. The properties manager also maintains an indicator of the type of network connectivity that the peer is likely to have to other servers. This is determined by looking at the characteristics of the network interface that the server has active at each hour. During the hours when the peer has slow network connectivity (e.g., the only interface active is a dial-up modem), the uptime cycle is marked to indicate that the machine has a low probability of being available.

6.2.5 The Data Manager

The data manager is responsible for keeping copies of the backup files on the local disk. The data manager maintains the time stamp when each file is backed up. The time stamp of the backup copy is computed according to the clock of the peer that contained the original copy. The data manager also maintains an index of all the files that are backed up on the local peer, along with its size.

Files that are backed up by using incremental differences from the older versions may be stored locally with a listing of the incremental changes rather than the full version of the files. The data manager is responsible for managing the differences and delivering the complete file to any requesting user.

The data manager keeps all files in a compressed format to reduce its storage requirements. When a file to be archived is encountered for the first time, the data manager also checks to see whether the file is identical to another file of the same name from another system. If the file is identical, then only the metadata (time stamp, owning node, etc.) information is created for the new file, with a pointer made to the copy of the existing file.

Most enterprises maintain a limited document retention policy in order to contain the amount of storage needed at the backup site. Thus files would typically not be backed up beyond a period of a few years. Some types of data, for example, billing records, accounts records, and tax-related information, are maintained for a larger period of time, for example, 7 years, depending on government regulations and operating practices. The data manager maintains the time period for which a file needs to be maintained in the metadata, and files that are past their document retention period are eliminated from the backup.

6.2.6 The Schedule Manager

The schedule manager is responsible for scheduling automatic backup of the contents of the local file system at regular intervals. The scheduling of backup operation may be done by using the facilities provided by the operating system.

Most operating systems provide for mechanisms that allow the operation of a task at regular schedules. The cron program available on most Unix systems provides for this capability. Versions of the Windows operating system subsequent to Windows-NT provide for the ability to run tasks on a schedule. On most servers, it has been traditional to use such facilities to schedule backup of server data at a time when the network is likely to be idle, for example, between midnight and 6 am. On systems that are continuously operational, this approach works quite well.

The scheduled backup system works in a similar manner on personal computers and laptops. However, such systems are shut down more frequently and may not be regularly up at specific times of the day. The schedule for the backup of such systems is usually launched when the system is booted up. To avoid any synchronization across backups of different client systems, most programs would impose a random delay after the system is booted before the backup process starts.

The scheduler on a peer-to-peer backup service would operate in a similar fashion, using the operating systems services to launch itself and to invoke the backup operation at random periodic intervals.

6.3 SECURITY ISSUES

One of the issues in creating backup data at different peers is that of security. Some of the files present on a user's workstation are of a secure nature and should not be visible to other users. However, backing up the files to another peer may make some of the sensitive data become visible to another user. This vulnerability is a unique problem in the peer-to-peer approach, because a centralized backup system usually is built so as to provide isolation among the different users.

To address the security concerns, a peer-to-peer system may encrypt a file being copied over for backup by using a key known only to the original user of the machine. To encrypt a file, an independent secret key for encryption is generated for each file. The key generation is done by a deterministic algorithm that takes four inputs: (i) a password specified by the user, (ii) the identity of the system where the file is created, (iii) the name of the file being backed up, and (iv) the time stamp on the local system when the copy is being made. When a backup copy is made, the peer with the backup copy has the information about all the parameters (except the first one) as the metadata maintained in its data manager. Thus the originating host, which knows the first parameter, can regenerate the key for the file during restoration period and retrieve the file. The algorithm for key generation can be made so as to generate strong cryptographic keys. A typical key generation algorithm could take the concatenation of the four input parameters, compute the exponent of the concatenated binary number, and then take the result modulo as a maximum key size as the secret key for encryption. Any secret key-based encryption scheme can then be used to encrypt the file.

The one drawback to this approach is that a user forgetting his/her password for backup in the system will have no way to recover or reset the lost password. To assist them in recovering the password, the system maintains a known text file that is encrypted and stored locally. These two files are not copied to the other peers. When a user forgets his/her password, the system provides utility tools that would help recover the password by trying to compare different possible combinations against the known file. The approach would be time-consuming and would only work if the known file is not corrupted on the disk.

Another issue with user passwords is that the password can not be changed, because that would not match with the keys used for backing up the previous files. To address that issue, the data management system maintains a local file containing all of the previous passwords and the time duration they are used.

This local file is always encrypted with the current key and backed up to other machines with the standard replication mechanism. When a user changes the password, the old password is appended to the password file and the password file reencrypted with the new key that would be generated by using the new password.

6.4 HYBRID DATA MANAGEMENT APPROACH

The approach to handling security and metadata management in peer-to-peer systems is not as elegant as it is in centralized backup systems. Centralized backup systems permit a easy way to manage user passwords, allowing users to reset them as needed, without the need to maintain a history of older passwords used. A good way to obtain the advantages of centralized security management and the cost savings of a peer-to-peer data backup system would be to use a hybrid approach.

In the hybrid approach, the metadata associated with each file are maintained at a central repository while the actual files are maintained at peers. The central repository is also responsible for managing the passwords of the users. The central repository uses a public key cryptographic approach to encrypting the files on the peers. All files are encrypted with the public key of the repository. Thus the issues of key management are restricted simply to managing the keys of repository.

The repository also maintains an index of the peers that have the backup copy of the different files. Thus the repository has a database that can be quickly searched to determine the set of peers that may have a copy available. The restoration process can be done much faster because the peers containing the backup are now easily identified and can be located without a broadcast query search on the peer-to-peer overlay.

All backup and restoration requests are made to the repository. The repository authenticates the identity of the user making the request and searches for a suitable set of peers for the machine. The repository identifies the peers that will make the backup copy, and the files are copied directly between the peers. After the successful backing up of a file, the metadata information at the repository is updated.

The hybrid approach requires a central repository, although

this repository would require much less bandwidth and data storage than a full-fledged centralized backup solution.

6.5 FEASIBILITY OF PEER-TO-PEER DATA BACKUP SERVICE

In this section, we look at the circumstances under which peer-to-peer backup service would be a viable alternative to the traditional server-based backup service. The architecture and design of the system require that the peers engaged in the data backup service have spare disk space that can be shared and allocated for the backup of data from other peers.

The good news is that most of the personal computers, as well as the laptops, operational in an enterprise have a large disk space that is often sparsely used. Industrial surveys indicate that up to 60% of storage capacity in an enterprise typically remains unused [37]. With the recent advances in the capacity of the disks on PCs and laptops, the amount of unused storage on the clients is likely to be much higher than the above figure. The excess storage capacity available in the devices can be exploited to provide data backup services to other devices in the enterprise.

The bad news is that client workstations are not as reliable as the backup server, and the key challenge in a peer-to-peer approach would be to ensure that the collection of several less reliable machines results in a highly reliable system from which a backup copy is almost always available for restoration. As a result, multiple copies of a file must made on the other peers so that the backup copy may be available with a high degree of probability.

The two factors, combined together, imply that the peer-to-peer data backup service would only work in environments in which the average personal computer disk is relatively empty. As a rule of thumb, if three backup copies must be made for a file in order for it to be backed up, the average PC disk should be less than 25% utilized in an environment. If disk capacities continue to increase the way they have been increasing, and the workload in an office environment remains geared toward the current mix of largely text-based files, then peer-to-peer backup systems could offer a viable solution to the data management in most enterprise environments.

On the other hand, we are also seeing the emergence of applications that consume a large amount of disk space. Digital pictures and digital video clips create large files, and environments that are geared toward multimedia content would easily find most PCs running into a shortage of disk capacity. A peer-to-peer backup service may not be appropriate for such environments because there will not be enough space on the computers to support its operation.

The focus of this chapter has been on pure peer-to-peer systems in which no backup servers are involved. Although we have examined the backup from the perspective of an enterprise, similar peer-to-peer techniques can also be used for personal area networks [39]. Peer-to-peer techniques can also be used to link multiple traditional backup servers together so that they work like a much more scalable backup server. In such an arrangement, the backup servers copy or retrieve data from other backup servers as needed. Some scalable systems such as TESS [38] implement such a peer-to-peer approach for linking backup servers.

7

PEER-TO-PEER DIRECTORY SYSTEM

A directory is an application that stores information about other entities that could be needed by a computer system. As an example, a directory would store information about the printers available within a building and their properties. Computer programs could use that information to find the right way to communicate with the printer in order to print files properly. As another example, most enterprises maintain a directory listing of their employees, which enumerates information about the employees such as their E-mail addresses, departments, and phone numbers. The directory can be used by other employees to communicate with their colleagues, as well as by computer applications to determine the appropriate privileges available to any specific employee, such privileges depending on the country and department in which the employee works.

The typical implementation of a directory follows the client-server model. The directory server contains a database of elements, and the clients access the server with an access protocol such as Lightweight Directory Access Protocol (LDAP). The directory server model works well when information about a large number of entities needs to be stored at a well-known location. However, the creation of the directory server requires centralized control and maintenance of the server and requires the updating

Legitimate Applications of Peer-to-Peer Networks, by Dinesh C. Verma

ISBN 0-471-46369-8

of information stored at the server. In applications in which the clients need to publish resources available locally, as well as access information about resources available at other machines, a peer-to-peer implementation of directory systems could provide a lower-maintenance and more scalable solution.

In this chapter, we first look at the structure of a directory server. This is followed by a discussion of the way in which a distributed peer-to-peer implementation of a directory system can be done. Finally, we discuss some of the other usage scenarios of distributed peer-to-peer directory systems.

7.1 LDAP DIRECTORY SERVERS

A directory server would typically consist of two components, a database of entries and an access protocol that is used by clients to access the database of entries. The database follows a hierarchical structure for naming the entries, and the predominant client access protocol is LDAP [40].

The database is organized as consisting of many types of objects, each type defined by a set of attributes. The types of objects and the attributes that are included in each object are defined by the directory schema. The schema for a directory can be customized for different types of applications. Two attributes are mandatory for each instance of an object—a distinguished name that is a unique identifier given to the object and the type of the object. The distinguished names arrange all the objects in a tree hierarchy, the hierarchy defining the name space for the directory. In this tree hierarchy, each object has a distinguished name that is obtained by adding a suffix (or prefix in some directory implementations) to the distinguished name of its parent in the hierarchy. Each directory has an object that forms the root of its name space, which will have the shortest distinguished name of all the objects in the hierarchy.

As an example, a simple directory of employees in an organization can consist of two types of objects, department objects and employee objects. A department would have attributes like its name, its manager, its function, etc., and an employee would have attributes like name, telephone number, etc. A department would have other departments and employees as its children in the name space tree, each child of type department would be a subdepartment, and each child of type employee would be a person who works in that department. The distinguished name of each entry would be formed by adding a new suffix to the distinguished name of the parent entry. The distinguished name of the root entry for the directory for XYZ Corporation would be "o=XYZ." It may have two department children with distinguished names like "o=XYZ, ou=Marketing" and "o=XYZ, ou=Development" and an entry for each of its employees with distinguished names like "o=XYZ, cn=Mike." An employee of the marketing department could have a distinguished name like "o=XYZ, ou=Marketing, cn=Fred." The abbreviations o, ou, and cn are used for organization, organization unit, and common name, respectively, but any set of abbreviations appropriate for the usage environment can be used in most directories.

The access protocol LDAP provides for the set of operations that a client can perform on the server. The operations provided by the protocol are (i) lookup of an entry given its distinguished name, (ii) create a new entry given its distinguished name and set of attributes, (iii) modify an existing entry given its distinguished name and a new set of attributes, (iv) delete an existing entry given its distinguished name, and (v) search for entries matching a search criteria under the name space tree below a given distinguished name, in addition to operations that allow a client to connect and disconnect from the server. Each of the operations (except for the connection and disconnection) needs a distinguished name of an entry as one of its arguments.

Multiple-LDAP directory servers can cooperate with each other to provide an aggregated directory server. This cooperation works in one of two ways. If a client connects to a server that does not have a distinguished name for the operation being performed, it can send a redirection request to the client. Alternatively, it can act as a proxy for the client and obtain the results of the operation from the server that has the appropriate entry. The set of cooperating servers and the distinguished names of the root entry they contain is provided as part of the server's configuration. Multiple-directory servers can also replicate the contents of each other's directories with standard protocols [41].

The above discussion provides only an overview of the basic functions of the directory server. The details of more advanced features of LDAP, as well as a more detailed overview of LDAP, can be found in several books such as [40], [42], and [43].

7.2 WHY USE PEER-TO-PEER DIRECTORIES?

In contrast to the traditional directory server, the peer-to-peer directory server would consist of all the entities involved in a directory system working together to provide the information about their set of objects and entries to each other. Thus there is no client accessing a server with a stored set of information, but all the information in the directory is replicated across the various cooperating clients. Compared with the standard server-based directories, a peer-to-peer directory server would offer the advantages of lower administrative overhead, better scalability, and self-management.

If the information that is stored at the various clients is available locally at the various clients, it is easier to maintain them locally. The different systems can then cooperate to distribute the information among the different members, and there is no need for a centralized administration of the information. Thus a peer-to-peer directory structure would tend to be easier to administer and update with local information.

Similarly, if the information can be maintained at many different sites, then a much larger set of information can be maintained within the directory compared with a single-directory server. Peer-to-peer implementation can leverage the power of many machines to provide for a much more scalable solution than that of a directory server.

Peer-to-peer technologies enable a system of multiple directories to manage and configure themselves automatically. They let each of the directories learn about the type of entries stored in the other directories and thereby enable client requests to be redirected to the right directory automatically without the need for any manual intervention.

In some contexts, peer-to-peer directories help alleviate the problem of having multiple administrative domains. If two or more corporations want to cooperate and provide their information in their own administered directories to appear as a single common directory, they can use a peer-to-peer approach to obtain this objective. Each of the directories is a peer to the other directories, and together they provide the abstraction of a combined directory service to their clients.

This last example shows the use of peer-to-peer techniques to build a directory server that can be accessed by clients as if it is a single unified directory. These directories are a way of building a scalable set of servers by combining many servers together. Another way to coordinate multiple servers is by defining a hierarchical relationship among them. An example of such a hierarchical relationship would be to put a front-end proxy in front of multiple-directory servers or to designate one of the directories as the primary that redirects users to the appropriate other directory. Another example would be to arrange the directory servers in a tree hierarchy as in the Internet domain name service. In a peer-to-peer system, all of the directories will be equivalent without a hierarchical relationship with each other. This enables them all to act as equals without any undue importance given to any one of the directories.

7.3 A PEER-TO-PEER DIRECTORY SYSTEM

In this section, we describe the design of a peer-to-peer implementation of a directory system that allows directory information to be stored into many different nodes. The directory consists of records that are contained and managed within all the nodes that are established within the entire system. The records must conform to a schema, and each schema creates an independent instance of the peer-to-peer directory system. To build a peer-to-peer directory system, the following components are needed:

A schema maintenance component: This component is responsible for distributing the schema to different participants and for updating the schema when requests are made.

The operation processing component: This component is responsible for finding the right peer or set of peers that can process any of the specified directory operations and obtaining the results by combining the information returned from all of the peers.

The access management component: This component is responsible for allowing updates and queries to entries in a directory to only those peers that are entitled to store and represent their own information.

These components are described in further detail in the subsections below.

7.3.1 Schema Maintenance

The schema in a centralized directory system is created as part of the directory's configuration. The schema of a directory service is used to check for the validity of the attributes and objects that are stored within the directory. In a distributed peer-to-peer system, this configuration information can be created by an administrator. However, the key issue confronting schema maintenance in a distributed peer-to-peer environment would be managing the distribution of the schema with its different versions and preventing updates from multiple possible administrators from the site.

In most implementations of a directory system, we would expect the schema for the directory service to be determined in advance and not change subsequently. The simplest approach for schema maintenance would be to include the schema as part of the software that each peer must install. If the schema is never changed subsequently, the initial installed software includes the schema that is supported within the system.

An alternative approach is to have the schema not be included in the software install but defined subsequently by an administrator. An administrator of the system would define the schema to be used by the system and then broadcast it to all of the participants in the system by using any of the techniques described in Chapter 2. The participants receive and use the distributed schema to check the structure of objects that are stored within the directory. A peer can only participate in the directory system after it receives one version of the schema.

During the operation of a directory system, the schema may need to be updated. The schema upgrade must be done so as to be backward-compatible with the existing entries within the directory. An example of such an upgrade would be to define a new type of object that can be added to the directory entries or to include a new optional attribute in existing objects. Note that some of the changes in the schema may invalidate the existing records in a directory. An example of such a change would be the introduction of a mandatory attribute in an existing type of object/record, which would make the existing entries invalid. The introduction of such changes would require the deletion and then recreation of many existing records to ensure proper operation in the future.

When a schema update is made, a version number is assigned to each of the versions of the schema. When a schema change is broadcast, each peer accepts the latest version of the schema and then uses it to update its local configuration. In a peer-to-peer system, each peer would periodically query neighboring peers to see whether they might have a schema version that is more recent than the one it has and, if so, retrieve a copy and update the local version.

An improper schema version can readily bring down the peer-to-peer system, and therefore each new schema version received at a peer must be checked to ensure that it is correct and does not contain any conflicts with existing records. If a schema appears to be incorrect or has a syntactical error or other types of inconsistencies, it should be ignored and a error message sent out to the administrator of the system (or the entity making the change) about the errors in the new schema version.

7.3.2 Operation Processing

Operation processing refers to the task of performing an operation (create, modify, delete, search, or lookup) with the set of directories that make up a system. We assume that all of these operations can be initiated by any of the peers and should change/obtain the records from all of the other peers participating in the directory system.

To understand the different options available for implementing a peer-to-peer directory system, let us divide the various operations into two categories—the read operations and the write operations. The read operations do not modify the records within the directory system, whereas the write operations do modify the records within the directory system. In the set of standard operations, lookup and search are read operations and the others are write operations. One of the key aspects of the write operations is that that they only operate on one single record of the directory at a time, namely, the one that has the specific distinguished name specified within its argument. Given that only one entry with a specific distinguished name can be created in the directory system, the operation must be routed to the peer that has the right distinguished name and then be performed there. The lookup operation also reads only one specific record and thus can be performed like the write operations. The search operation needs to find all records that are underneath a specified distinguished name in the name space and thus would need to be performed in a different manner than the other operations. Thus all operations in the directory system can be classified into a write operation of a single record, a read operation of a single record, or a search operation across multiple records.

To process any of these operations, four issues must be taken into account: (i) how the different directory records are distributed throughout the different peers, (ii) how to route the operation request to the right peer that can perform that operation, (iii) how to combine the response to an operation from all the peers, and (iv) how to process the operations that access the same entry into a sequential manner so that the behavior of the system remains consistent from an external perspective. All four of these issues are intertwined with each other, and various alternatives are available to implement them.

7.3.2.1 Local Placement of Records. The first approach that one could use to build a peer-to-peer directory system is to impose no restrictions on how the records of the directory system are stored at the various peers. In this model, each peer can create any record in the directory system locally. As mentioned above, the records created in a directory system must have a unique distinguished name. A create operation can be performed by the originating peer creating a local record after checking that the assigned distinguished name does not exist in any of the other peers. The check can be performed by doing a broadcast of the proposed distinguished name to the other peers, who would respond back with a conflict message if they have an existing record with the same distinguished name. If no conflicts are reported within an elapsed time period, the peer would go ahead and create an record with the specific distinguished name.

The duplicate detection process may fail to detect duplicates if the peer with a conflicting distinguished name is temporarily unavailable. To minimize the probability of this happening, each peer can maintain a cache of the recent requests for names that it has seen, as well as a set of all names that each of its neighboring peers in an overlay has maintained. The size of this cache can be predetermined to limit the amount of disk space usage. In that case, the duplicate detection process would work as long as one of the peers that knows of the existence of the name is up and available on the system.

The duplicate detection mechanism is probabilistic, and there is a finite chance that duplicates may still be created. Other cases in which duplicates may not be detected completely can arise because of errors in the underlying broadcast mechanisms of the overlay network, leading to lost requests or responses. Therefore, a peer-to-peer implementation of directories would need to account for mechanisms that discover the presence of duplicates of its records occasionally. One mechanism would be to periodically perform a lookup of distinguished names that are created locally. If duplicates have been created because of a race condition, they would be identified and corrective actions can be taken to remove one of the elements. One scheme to use for determining the element to be removed will be to associate a unique identifier with each record (e.g., a concatenation of its creation time and the address of the creating peer). The record with the large unique identifier value would be deleted.

All of the other operations are broadcast to the peers. For the read/write operations of a single record, exactly one of the peers would have the entry that matches the specified distinguished name. That peer can perform the required operation and send a response back to the originating peer. If more than one peer responds back, then the originating peer can detect a conflict in distinguished names and notify all the respondents to resolve the conflict in the distinguished name space. In the search operation, each of the peers will respond back to the originating peer if they have a nonzero number of records that match the search criteria. The originating peer can combine all of the responses together to obtain the results of the search.

The local placement of records is a simple way to implement the directory search for peer-to-peer applications. However, this mechanism is relatively inefficient in terms of its bandwidth usage and search latency. The need to coordinate across all of the nodes before creating a record also results in a long latency in the creation of the records within the directory.

One way to make the record creation process much faster would be to use a central registration server approach for determining whether a distinguished name has already been used. In that case, each peer checks with the registration server to check for the uniqueness of the distinguished name before it performs the create operation and unregisters the name when that record is deleted. The existence of such a server can make performance of the task of creating and coordinating new entries much faster.

7.3.2.2 Name Space Partitioning. A much more efficient implementation of a peer-to-peer directory system can be performed if each peer maintains an independent portion of the name space imposed by the directory server. In this context, each peer is assigned one or more distinguished names, and the peer is responsible for storing all of the distinguished names that are present below that distinguished name in the name space. An operation only needs to be routed to the responsible peer for the distinguished name that is specified as the argument in the operation, and the peer can then perform the appropriate operation and return the results.

If all operations only require manipulating a single record identified by its distinguished name, the distinguished name can be used as the key for the record and the operation can be considered an extension of the file storage/retrieval application described in Chapter 5. However, the search operation in the directory operations requires traversal of a hierarchical tree in the name space and access of many records, each of which has its own distinguished name. Because keys obtained by distributed hashing functions need not preserve the hierarchical relationships in the name space, many of the techniques used for maintaining distributed hash tables in Chapter 5 are not applicable to directories.

A focused search technique, which maintains a routing table to forward queries selectively along certain preferred nodes instead of forwarding queries to all of the neighbors, can be used to build efficient peer-to-peer implementations of directory services. Each node takes ownership of a part of the name space. The part of name space it claims ownership to may be determined by local configuration or be based on some heuristic allocation. Each node then broadcasts information about the name space of which it has ownership to the other nodes. This information is then used by the other nodes to build a routing table and to send the operation over to the responsible node with the routing table.

For the sake of illustration, let us make the assumption that each of the participating peers has enough disk space to maintain the distinguished name of each record within the entire directory system, as well as a pointer to the peer that is maintaining the record identified by that distinguished name. In some contexts, this assumption may indeed be valid, although this may impose unrealistic disk space requirements in many other instances. Assuming that the disk space required is feasible in all the peers, each originating peer can identify the other peers to which any specific operations ought to be routed. The originating peer can then redirect the operation to the right peer. The routing information can be propagated relatively easily. Each peer creates the routing information for its own node and sends it to all of its neighbors. Each of the neighbors includes the information about the various entries and sends the consolidated update to its own neighbors. The consolidated set would consist of the distinguished name of all records that a peer recently received updates about plus all the records about which it already had maintained existing information. After a period of learning, each peer in the system would know exactly where each record is located.

Because the disk space required in such a scheme is excessive, we should look at a mechanism by which it can be reduced. One way to improve the efficiency of the system would be for each peer to publicize only the distinguished name of the root of all of the records that it has in its local name space. Each of the originating peers can then simply compare the distinguished name specified in any operation with that in the routing table and forward the request to the peers that have matching prefixes in the routing table. Because one would typically expect thousands of directory entries at each of the peers, this mechanism would reduce the total routing table size tremendously.

The routing table size may still exceed the maximum disk space a peer may wish to devote for the purpose of running the directory application. This implies that the consolidation process for the routing table information received from the neighbors must be truncated in an intelligent manner. Each routing table information contains the information about a distinguished name and a peer that contains the record with that distinguished name. If the overlay links connecting the various neighbors in a peer-to-peer system are relatively fixed, then the address of the peer can be replaced with the address of the neighbor that can be used to reach the peer. Furthermore, the node can compute the longest common prefix (or prefixes) that specifies that the same neighbor can use that information to route the requests to the peer that is likely to contain that record.

One way to consolidate the name space information without any loss of information is to traverse the combined set of entries obtained from all of the neighbors using a breadth-first traversal scheme. Some of the distinguished names will be cut off when the traversal exceeds the maximum limit placed on the size of the routing table. However, records of that name would be under the name space of at least one of the records that are still included in the routing table. This ensures that each operation can be routed to the node that contains the record in a series of hops.

Thus an efficient distributed directory system can be made by creating a routing table that includes application-level information. This is an extension of the routing protocols used within the Internet, which only look at network address information. The current extension accounts for the name space of the directory service to guide how requests are routed in the overlay network that defines a peer-to-peer infrastructure.

7.3.3 Access Management

In any directory system, access management must determine whether a specific peer should be able to view or modify the contents of a specific record. In a distributed infrastructure, the directory server needs to address the same concerns. An access control list must be assigned with each of the entries that defines who can and cannot perform specific operations within the directory system.

In the client-server model of directory systems, such access management lists are usually provided as configuration information for the system. To that extent, access control lists can be distributed like the schema files, as described in Section 7.3.1. However, the access control lists changes much more frequently than the schema files.

As in other peer-to-peer applications, one can reuse the capabilities provided by public key cryptography to enforce access control among the different entities. Each peer can be provided with a certificate that acts that proof of its identity. If an anonymous use of directories is desired, a certificate is used that identifies the role of a peer. A role is an alias for a peer and is the entity used to define access (read/write) privileges to the directory entry. All certificates are signed by an entity that is trusted by all the peers and whose public key is provided as part of the initial configuration information to the peers. On each operation, the participating peer would check the access privileges before performing the desired operation on the entries contained within the directory.

Access control schemes based on public key cryptography work well when all peers are working properly but can fail in the presence of a malicious peer when name space partitioning techniques are used to distribute the records of a directory. Because the records are now present on peers that are not the original creator of the content, the records can be read by the peer and passed on to unauthorized recipients. One possible solution against colluding peers is to encrypt the directory records. The encryption is done with a key that is only available to the authorized users. An authorized user can communicate with the original creator of a record to authenticate himself and to obtain the secret key. Other models for encrypting the contents of a directory, and for distributing the right keys to the authorized users, can also be used.

7.4 EXAMPLE APPLICATIONS OF PEER-TO-PEER DIRECTORY

In this section, we look at some peer-to-peer applications that can leverage distributed directory services. Let us consider a peer-to-peer system that is intended to enable the participants to chat with each other. For simplicity, let us assume that the chat application only involves communication between two participants. A typical implementation of the chat program consists of each of the two participants talking to the chat server and the chat server providing the information to be displayed on both of the participants' screens. The chat server also acts as the meeting point where the two participants can find who else is online.

Chat is a very useful application, but it requires the establishment, maintenance, and operation of a server. Such servers are operated for instant messaging services offered by many sites and internet service providers. However, running these servers requires dedicated machines and staff, which can add up to a significant cost. We could envision a system in which the participants want to have a chat application without the need of an intervening chat server. As long as the participants are connected to the same network (e.g., the Internet), they could use the software to chat with each other without needing such a server.

Such a chat application would begin by implementing a directory service in a distributed peer-to-peer fashion. Each user of the application obtains a unique name for in the directory system. The unique name is an online alias that is maintained uniquely by means of the schemes described in Section 7.3.2.1. The chat process itself consists of two steps—looking up the directory system to determine the current coordinates of a user (such coordinates would include the current IP address and port number that the chat application of the user is running on) and then establishing a connection to the appropriate machine to initiate the conversation. When a user comes online, its chat application updates its record in the directory system with the current IP address and port that it is running on.

The above example application does not require a coordinated name space across the different sets of applications. However, a coordinated name space would be needed when several peers are collaborating to provide services to each other. As an example, let us consider a system of collaborating peers on the Internet that need to convert text documents from one format to another. Most users are familiar with the common formats that text documents take, for example, Microsoft Word format, Lotus Wordpro format, Adobe pdf format, Postscript format, etc. A wide variety of programs exist that enable conversion of a document from one format to another.

A directory format can be created where the records are created in a hierarchical name space with the first level of hierarchy being defined by the format of the original document and the second level of hierarchy being defined by the format of the new document. The final distinguished name is a unique identifier of the participating peer. Thus a peer that is running an application capable of converting a Postscript file to a pdf file would be creating a record for its service with a distinguished name like "o=global, in=ps, out=pdf, id=576," where the last field is unique across all participants. A unique id for this application can be created by selecting a large random number, where each digit is determined randomly by the computer. If a 15-digit decimal number is generated randomly, the chance of two peers obtaining the same number is less than one in a quadrillion and almost unlikely to happen with a million or so participating peers. Each directory entry contains the information needed to enable a remote client to invoke the conversion program available at the server.

When a conversion service is needed, each of the participating peers can issue a directory search to find a set of nodes with software available to perform the right type of conversion and invoke its services.

8

PUBLISH-SUBSCRIBE MIDDLEWARE

Publish-subscribe systems are a niche area in building enterprise systems that form an important aspect in integrating and developing applications in many businesses. Publish-subscribe middleware enables the publishing of information from one set of applications and its dissemination to another set of applications that are interested in obtaining the associated information. Many applications of publish-subscribe systems arise in the financial and news industries. Many trading companies obtain current stock price quotes from the stock exchange and need to disseminate them to the different applications that analyze the quotes to make decisions about buying or selling that stock. Organizations involved in the business of publishing news also need to connect their news producers, the reporters in the field, to other staff that would edit, format, and prioritize the news and then send the revised, consolidated reports out to the public outlet of the news organization, for example, the website used to publish the news to the external world. In most of these applications, the producers of information need to be connected to the consumers of the information.

Software that provides connectivity between producers and consumers of information is called publish-subscribe middleware. Such middleware is commonly implemented with a client-server approach but can be implemented equally well with a peer-to-peer

Legitimate Applications of Peer-to-Peer Networks, by Dinesh C. Verma

ISBN 0-471-46369-8

approach. This chapter provides an overview of publish-subscribe systems and presents a common design for the client-server approach as well as the peer-to-peer approach for building these systems. Finally, we present a comparison of the two approaches and provide some examples of applications where the peer-to-peer approach for publish-subscribe systems would be suitable.

8.1 OVERVIEW OF PUBLISH-SUBSCRIBE SYSTEMS

Publish-subscribe technologies typically use a subject or topic name as the mechanism to link publishers and subscribers of information. Publishers produce messages on a particular subject or topic name, and subscribers register their interest in a specific set of subjects. Once an application registers on a subject topic, it receives the messages that are created by the publishers on that topic. Information is pushed to subscribing applications as it is generated. Publishers and subscribers can join and leave at any time. The middleware is responsible for routing messages between the publishers and the subscribers.

A typical use of publish-subscribe systems is in the dissemination of stock quotes. Information dissemination companies such as Reuters and Bloomberg provide the software that consolidates the current trading information in the stock market and provides the quotes to subscribing customers, typically enterprises engaged in stock trading or brokerage operations. Within a trading company, quotes are received by an application that interfaces with the information dissemination company and then relays it to the other applications within the company that analyze the quotes and take action based on the current quote information. The trading company has a publish-subscribe system with only one publisher—the application that interfaces with Reuters or Bloomberg. The name of the stock involved in the quote provides the subject of the information published in the system. Applications receive the quotes for specific stocks and act on them.

The information dissemination company can also use a publish-subscribe system to manage its customers. The subject or topic name in this publish-subscribe system could be any categorization of quotes. One categorization could be based on the stock exchange where a trade occurred, for example, quotes on all trades happening in the New York Stock Exchange versus the American

Stock Exchange. Another type of categorization could be on a bundle of trading companies, for example, all trades of the 30 companies used for computing the Dow Jones Industrial Average, the trades of the companies used for computing the Standard & Poor's S&P 500 index, or the trades of the companies used for computing the S&P 1000 index. The publishers of the information are systems that interface with the stock exchange's computers, and the consumers of the information are the enterprises that are customers of the information dissemination company.

Publish-subscribe middleware provides the infrastructure on which the information dissemination applications can be built. All information exchanged in such systems consists of messages, and each message is associated with one or more subjects.

Publish-subscribe systems can be implemented in two broad ways, the first one using broadcast to all of the subscribers and the second one consisting of creating intermediary message routing systems. In the first category, all messages are broadcast to all of the recipients and software at the recipient filters out the messages that are not of interest to the recipient, that is, do not belong to a topic that the recipient has registered an interest in. The second category of middleware creates and establishes a routing path for each subject topic, and messages belonging to a specific subject are sent to recipients among the broadcast tree.

8.2 SERVER-CENTRIC PUBLISH-SUBSCRIBE SERVICES

A traditional publish-subscribe system is implemented by following a client-server architecture in which a messaging client is associated with each of the publishers and subscribers. The messaging clients register the topics for which they would like to receive messages with one of the messaging servers. The messaging servers are responsible for communicating with each other and relaying the messages to the other messaging servers. The communication between the various messaging servers is effectively peer-to-peer in philosophy, with the difference being that it is often statically configured and does not use the dynamic self-configuration that is a characteristic of large-scale peer-to-peer systems.

The client-server architecture for publish-subscribe systems is geared toward a design point that consists of a few highly scalable message servers handling a large number of clients. Such a system can handle a high volume of messages for transmission within the system with a low number of messaging servers. When the messaging servers are moderate in number, they could be configured by hand to communicate with the other servers and in effect set up the overlay connecting them to each other manually.

As mentioned above, the client-server architecture for publish-subscribe systems can follow an approach of broadcasting or building a message routing system. When broadcasting schemes are used for sending messages, all of the messaging servers are connected together in an overlay. As an example, the messaging servers could be configured into a tree topology spanning all of the messaging servers. The messaging client on each publisher sends the message (with its associated subject) to one of the messaging servers, which then relays it to all of the other messaging servers. Each receiving messaging server checks the subject header to see whether it is of interest to any of the messaging clients that are registered with it and, if so, forwards the message to that client.

As long as the number of servers is small, this approach can work quite well. Checking the subject of a message against clients that have registered can be done fairly efficiently by setting up appropriate data structures at the receiving clients. By creating an index that maps each subject/topic to the set of messaging clients that have subscribed to that topic, a messaging server can quickly determine to which of the clients a message ought to be forwarded. Some messaging systems allow subscription on a group of topics with wild cards. In those cases, an index can be set up if the messaging server knows the overall set of topics that need to be advertised on. If message subjects can be arbitrary and not known in advance, the server can use the somewhat slower scheme of checking the subscriptions of each client against the subject of each incoming message.

The broadcasting approach has the disadvantage that some messages may be forwarded to all messaging servers. A messaging server would receive messages even on subjects for which there are no registered clients with an interest. If it is highly likely that a message is of interest to at least one client registered at all the messaging servers, this may not be a large inefficiency. However, when the number of subject topics is large, this scheme can be rather inefficient.

The message routing approach within messaging servers alleviates that scalability problem. It builds a distribution tree on each of the subject topics that is used to distribute messages only to those sets of servers that have clients registered for that subject. No messaging server receives redundant messages that are not relevant to it, but the middleware has to do more work to establish a subject topic. For each new topic that is introduced into the system, the distribution tree corresponding to that topic must be established.

A messaging system can also take a hybrid approach in which distribution trees are established for some subset of the subject topics, but other topics are broadcast to all of the receivers.

8.3 PEER-TO-PEER PUBLISH-SUBSCRIBE SERVICES

There are many similarities between the communication used for distributing messages among the messaging servers in the traditional publish-subscribe middleware and the application-level multicast functions described in Chapter 3 of this book. In essence, the messaging servers in a traditional publish-subscribe middleware use application-level multicast to distribute messages between each other or use a broadcast scheme to reach all of the messaging servers. This implies that publish-subscribe systems can be readily built on a peer-to-peer infrastructure just as application-level multicast can be supported in a peer-to-peer environment.

Peer-to-peer implementation of a publish-subscribe system dispenses with the messaging servers and includes an instance of the messaging server with each instance of the messaging client. This increases the number of messaging servers that need to communicate with each other. The larger number of messaging servers would entail the use of self-configuring overlays and automated route creation methods as opposed to static configuration and manual administration of a small number of messaging servers. Furthermore, the nodes that are involved in the process of forwarding messages would tend to be much less reliable than the traditional case in which such functions are performed by messaging server nodes that would tend to be more reliable.

The peer-to-peer publish-subscribe mechanism could also use the same two basic paradigms that are used in traditional publish-subscribe middleware, namely, use a broadcast-based scheme or use a message routing scheme. Both of these schemes are described further in Sections 8.3.1 and 8.3.2.

8.3.1 Broadcast Scheme

In the broadcast scheme, each peer in the system handles registration requests from applications that are present on the local machine. It keeps a list of the local subscriptions containing information about the applications, the subject/topics that the applications are interested in, and the mechanism that can be used to send a message to the applications. The mechanism used to send messages to applications may involve sending a message on a queue from which the application reads, sending a message on a local connected system, writing an area of a shared file, or invoking a call-back function, depending on the implementation details of the middleware.

All the peers form an overlay that connects them all. The overlay could consist of a mesh of nodes, or it may take the form of a tree topology that interconnects all of the peers together. In the case of a tree topology, each node can send a copy of the message to the other nodes by sending it out on the links of the tree of which it is a member. Each of the other peers takes the incoming message and forwards it along all of the other links in the tree that it belongs to. The tree is a structure that does not consist of any loops and ensures that messages are transmitted to all of the receivers. Messages can be acknowledged by each of the nodes to ensure that they have been received by all of the elements within a node.

When a mesh is used for broadcasting, the loop-free property of the tree can no longer be assumed. In this case, each peer must ensure that it is not submitting duplicate copies of a message to a receiver. Elimination of duplicates can be done by associating a sequence number with each message. Each peer keeps a history of the packets that it has seen in the recent past and discards any duplicates that have already been seen in the recent past. In most common computing environments, keeping a history of message sequence numbers seen in the past few minutes would suffice to ensure that the duplicate messages have been eliminated.

An associated problem with messaging is ensuring that the messages are delivered reliably from the publishers to the subscribers. Part of the reliability can be obtained by having the communication at each hop in the overlay be reliable so that a message is never lost in transit. However, this in itself is not adequate to ensure end-to-end reliability if the underlying topology of the system is changing. Ensuring a reliable end-to-end delivery mechanism would require one of the different techniques for reliable multicast to be implemented at the peers. Such mechanisms are described in Chapter 3. These mechanisms consist of associating sequence numbers with the senders and message topics, with either the sender or receiver ensuring that the messages are acknowledged by all of the recipients.

As in the case of traditional networks, broadcast approaches work well when all the peers are likely to be the recipients of the messages. If different peers have interests in subscribing to different topics, the broadcast approach is not very attractive in a peer-to-peer environment.

8.3.2 Multicast Group Approach

A better alternative than the broadcast approach in a peer-to-peer environment is to build a separate application-level multicast group for each of the topics that are currently in use within the system. Each of the peers is connected to a set of other peers in the network, forming an overlay network. When a message is to be broadcast to all of the nodes, each peer would relay the message to all the links of an overlay except the link on which the message was received. For a subset of the subject topics, each node maintains information about a subset of the outgoing links on which messages belonging to a specific subject are forwarded.

The mechanisms by which the messaging system can be maintained are the same as discussed in Chapter 3 on application-level multicast. The different technologies described there can be combined to provide a system to set up routes for sending messages among an overlay created among all the participating nodes.

One such combination is the use of the broadcast and prune approach to create multicast communication groups. An underlying overlay connection is made among all of the peers that are participating in the system. Peers advertise the subject topics that are of interest to their neighbors on the overlay. A peer with only one neighbor will send the locally subscribed topics to its neighbor. A peer with multiple neighbors will combine its sets of locally subscribed topics with the sets of topics it receives from other neighbors. It creates a copy of subscriptions to be sent to all of the neighbors, with each neighbor receiving a copy of the local subscriptions combined with the subscriptions of the other neighbors. Each node will relay a message on a specific topic only to the neighbors who have explicitly registered interest in the topic. The scheme ensures transmission of the messages to subscribers after the subscription tree has been established, but it creates a potential for missing messages while the tree formation is in progress.

A complementary approach to the above process can be taken by building a process similar to that used in the Internet Group Management Protocol (IGMP). In this mechanism, each node broadcasts the messages to all of the outgoing links. However, leaf nodes (i.e., nodes with only one neighbor) that have no local subscriptions to a subject topic can request that messages not be sent to them for that specific topic. A node that receives notifications on a topic from all but one of its neighbors requests the other neighbor not to forward messages on that topic to it. Each node honors the request for some fixed time period, for example, 10 minutes, and a node has to renew its negative interest in a subscription topic when it receives a message on the above topic.

Whereas the first scheme described above builds a multicast tree one link at a time as subscribers come online, the second scheme prunes off branches from the broadcast tree as nodes without any subscribers are discovered. The former is better suited for an environment in which we expect a subject topic to be subscribed only by a few nodes; the latter is better suited for environments in which subject topics are subscribed by most of the nodes within the system.

8.4 COMPARISON OF APPROACHES

From a message forwarding perspective, the peer-to-peer approach is almost indistinguishable from the internal communication schemes used by the messaging servers themselves. The messaging servers use the same techniques for broadcasting and multicasting messages (reliably or otherwise) among themselves as the clients would do among themselves. From a technical perspective, the difference between the effectiveness of the peer-to-peer approach and the traditional approach is related to the issue of size. When a given number of application instances need to be connected by the publish-subscribe middleware, the peer-to-peer approach would result in a system that includes all of them in message forwarding whereas the traditional approach would result in a system that has a much smaller number of nodes participating in the forwarding process. As an example, if 1000 instances of an application need to be connected together for a publish-subscription service, a traditional system may create 10 messaging servers, each supporting 100 clients on the average. The traditional system would then hand-configure the topology of these 10 servers for forwarding messages and broadcasting to other servers. The peer-to-peer approach would require the forwarding of messages and broadcasting to all of the 1000 participants.

When implemented with a broadcasting paradigm for distributing messages, the traditional arrangement with 10 servers is likely to work more efficiently and reliably than the peer-to-peer mechanism. A peer-to-peer implementation, however, can be created so as to effectively create a hierarchy of nodes in two tiers, the first tier consisting of a smaller number of nodes that are involved in the message forwarding process and the other tier consisting of message senders/receivers that connect to the nodes in the first tier. This is a self-configuring structure that results in the same topology as that of the traditional network. The difference between the two would be that the peer-to-peer overlay would be self-forming and self-configuring. It would determine the right set of nodes that should become forwarding nodes based on the available computing power in each of the machines.

The traditional approach is likely to be more efficient from the point of view of the raw throughput that it can provide and the volume of messages it can handle through the system. This is because the messaging servers are dedicated to the task of message forwarding and matching subscribers to requested subject topics. In contrast, the peer-to-peer approach steals cycles from the normal functions performed by each machine to do the forwarding function. This puts it at a disadvantage compared with the traditional approach.

The traditional approach also has an advantage when it comes to tracking down the path of a message and diagnosing the causes of the problem when a message gets lost within the system. Because the path of a message from a publisher to a subscriber is effectively fixed by static configuration, a human can track the message through the system to determine whether or not it is being delivered to the appropriate subscriber. In the peer-to-peer infrastructure, messages take a more diverse set of paths and it would be harder to track down the cause when a message is lost in the system.

Although the traditional approach may be better on technical grounds, the peer-to-peer approach may be more appealing on financial grounds. The traditional approach relies on dedicated messaging servers. The presence of additional servers requires additional capital investment in procuring the server and, more importantly, requires a trained staff to operate and maintain the messaging servers. The peer-to-peer approach relies on using the hardware of existing machines. As a result, it can operate at a lower total cost of ownership, even if the solution is somewhat inferior in the degree of scalability it can achieve.

A peer-to-peer approach would be suitable for publish-subscribe systems that are needed to operate without a dedicated server, for example, in small and medium-sized businesses. It would also be suitable for environments in which there are a large number of potential participants, which makes the configuration and operation of messaging servers by hand difficult.

8.5 EXAMPLE APPLICATION

Almost all applications that use the traditional model of publish-subscribe systems can be reimplemented over a peer-to-peer implementation of the same. Earlier in this chapter, we considered the example of a stock quote distribution service. The stock quote distribution service could be implemented in the traditional manner but also on a peer-to-peer implementation as well.

There are some applications where a peer-to-peer mechanism for a publish-subscribe system can prove beneficial compared with the traditional approach. One such example is the pushing of information to several different clients on the basis of their preferences. Push-based technologies pioneered by companies such as Pointcast were popular in the late 1990s but slowly lost ground, primarily because of their inability to recuperate the heavy costs of operating push servers from their subscribers and advertisers. The traditional implementation of push technologies used a client-server approach, with clients periodically polling the servers to obtain new information of interest to the channels of information they had subscribed to. The servers needed to be highly scalable in order to meet the requirements of a large number of clients, and many push companies were unable to recuperate the cost of running the servers from the consumers and advertisers. A peer-to-peer approach using a publish-subscribe middleware would have been much more efficient in network bandwidth as well as costing less to operate.

The push technologies allowed users on different machines to select and subscribe news items that were of interest to them. Each topic that a user could select would be used as a subject/topic for a subscription to receive news articles or information related to that topic. Each of the different nodes present in the system would form an overlay connecting all the peers together. Most of the peers would be receivers of information, and a few peers would be publishers of the information. The publishers would use the overlay among peers to push data to subscribing peers.

For small files such as text files, the publisher would push the entire contents of the file into the system so that it reached the desired peer. For larger files, for example, video files or large graphics, the publisher would only publish information about where the file could be obtained. The publisher would make the file available for download with a web URI that subscribers could access. Peers who had an interest in the content would obtain a copy of the file to be displayed locally. Such peers could also provide additional sites from which the file could be downloaded. Each peer who downloaded the file would change the message flowing in the publish-subscribe system to include the fact that it was also available locally. New peers who had subscribed to that type of content could download the file from any of the multiple locations that it was available from. The information could be presented to the end user in the form of a screen saver that would be displayed on the computers when they are idle or in the form of a window displayed on the screen of active users.

9

COLLABORATIVE APPLICATIONS

Collaborative applications refers to a broad class of applications that enable people to communicate and work with each other. Some examples of such collaborative applications are instant messaging, which allows people to communicate with each other, as well as shared whiteboards and shared systems that enable people to store ideas and concepts and help each other solve the problems that they encounter in the course of their work.

Many collaborative applications exist today, and they are typically implemented using the client-server paradigm. The client-server approach provides for a much simpler implementation of collaborative applications, because the server can be relied on to solve issues related to synchronization and consistency. The peer-to-peer approach has to use more complex solutions to the same problem. However, the peer-to-peer approach eliminates the need for a server, and the elimination of the infrastructure can lead to substantial savings in the cost of running a collaborative system.

In this chapter, we look at some of the collaborative applications that can be provided by a set of peers working together without a server. We also look at another set of collaborative applications, ones that are not intended for people to work together, but instead are intended to enable people with a set of services that a single peer machine would not be able to provide on its own. We begin with a look at the general issues involved in the design of

ISBN 0-471-46369-8

various collaborative applications and then look at a variety of applications that can be implemented based on the peer-to-peer paradigm.

9.1 GENERAL ISSUES

A collaborative application consists of several people working together on their machines in a joint activity. The joint activity could be a conversation or chat happening among the participants, a shared document or image being worked on by the participants, or a shared database or set of documents that are accessed simultaneously by each of the participants. Depending on the nature of the collaboration, some key issues to be addressed are the following:

Order: When multiple people are performing actions independently, the actions performed by them need to be seen by each of the participants in the same sequence. In other words, a mechanism must exist for the participants to view all of the actions done collectively by them in the same manner.

Consistency: When multiple participants are able to view the same document or other shared set of data, the operations they perform on their shared document can sometime result in conflicts. Two participants may be updating the same line in a shared file with different text, or a participant may be editing an image that is concurrently deleted by another participant. The shared information must present a consistent view of the changes to each of the participants.

Logging: In some collaborative applications, it is important to maintain a record of the interactions that happen among the different participants. The logging may be needed for a variety of reasons, depending on the applications.

A client-server approach to implementing collaborative applications offers simple solutions to the above problems. Typically, the server is responsible for maintaining the shared information, and each of the participants gets to view a snapshot of the shared information. The order in which different requests arrive at the server can be used to order the actions of all the participants. Logging can be performed relatively easily by the server recording the different activities as perceived by it.

Consistency can similarly be implemented in a straightforward manner by the server acting as the intermediary. One way for the server to maintain consistency among the participants is to require each participant to obtain a lock on the shared information (or a portion of the shared information) before making an update. This ensures that no two participants can modify the same content at the same time. A more optimistic approach would allow the participants to modify the content without requiring the lock. Each update is sent to the server with the modified information as well as the original information that was the view of the shared information before modification by the participant. If the original information does not match the current information at the server, the update is rejected. Other approaches can also be developed that leverage the unique position of the server to simplify the task of managing consistency among various updates.

In a peer-to-peer implementation without a server, such issues are much harder to tackle. Ordering mechanisms are known for distributed environments [44], but they impose a lot of overhead over a wide area network. Consistency across distributed databases can be maintained by using a two-phase commit protocol [45]. However, performing a two-phase commit over a wide area network is not advisable from a performance perspective.

A practical solution to the issues in a distributed environment is to designate one of the many peers in the system as the moderator. The moderating peer acts as the server for the collaborative application. As such, the same approaches that are used for client-server implementation to resolve these issues can be used in peer-to-peer implementation as well. The only difference will be that the moderating peer could be any of the participants rather than a fixed central server. If the moderating peer leaves the application, another moderator would need to be chosen.

When there is a large set of shared information, more than one moderator can be selected from among the various peers. Each moderator is responsible for a different portion of the shared information. This distributes the load among the different moderators and can help in improving the concurrency of operations among the participants.

We illustrate the various approaches that can be used for a

peer-to-peer implementation by looking at various applications in the next few sections.

9.2 INSTANT MESSAGING

Instant messaging and conferencing is an application that has seen widespread growth among users of the Internet. Instant messaging allows participants in a session to type messages to each other that are displayed on their screens. Instant messaging is used widely among enterprise users as well as consumers to communicate with their colleagues and friends to exchange information with each other, make plans, and work together.

Instant messaging is typically implemented with a server that participants log onto. Each participant maintains a list of participants whose status (logged on/logged off) is shown on the screen. The participant can click on active participants to send messages to them or invite other active participants in an ongoing conversation with existing participants. Each message is sent to the server, which orders them in the sequence in which the message is received, logs them if necessary, and sends the message over to the other participants in the session to be displayed on their screens. The server thus acts as a broker coordinating among all the participants.

In a peer-to-peer implementation of the same service, the functions of the server must be performed by cooperation among all of the participants. The server is engaged in the following functions: (i) managing the identities of participants, (ii) tracking the status of participants, (iii) initiating sessions and managing the participants in a session, and (iv) serializing the messages sent by different participants into a consistent view that is sent to all of the participants.

With no server to perform these functions, the participating peers must take these responsibilities among themselves. The management of the identities of participants, ensuring unique identities and obtaining the current status of each of the participants, can be done by means of a distributed directory service implemented with the techniques described in Chapter 7. The peer-to-peer directory ensures a unique name for each participant as well as the ability to locate the information about that participant by other peers by means of a directory search operation.

When a peer activates the instant messaging application, there is no logging on to the server. However, the status of the peer is marked as being active within the local directory server. Each participant has a local address book of other peers with which it frequently interacts. At the start of the messaging application, each participant looks for the status of all the other participants in the local address book by searching for them in the directory. The status of the participants is then updated on the local screen. The status can be checked periodically and updated to reflect the current status as participants leave or join the system. Although the periodic check is not as efficient as checking the status from a server site, the additional overhead would not be significant for typical address books of less than a hundred entries.

A peer initiating a conversation with other participants can become the moderator of the session and act as the point that combines the incoming requests from all of the other participants and sends them out to the other peers for display. The messages typed by the participants are serialized in the order in which they are received by the moderator and sent to all of the other peers for display purposes.

In the client-server implementation of the conversation, the server acts as the repository where all of the conversations could be logged and recorded. The logging of conversation is desired in enterprise environments, for example, financial houses that would like to keep a record of any requests for financial transactions made with instant messaging software. In other environments, for example, for friends trying to establish a common meeting place for an evening coffee, such logging by the server may be unnecessary or even undesirable because of privacy concerns. Because a peer-to-peer implementation does not include the server as an additional party in a conversation, only the participants in the conversation can log its transactions. Any potential eavesdropper can be thwarted by encrypting the conversation.

When logging of messages is desired, the moderator of the session can record a log of the session. A session may be logged by any of the participants as long as the session is given a unique identity. Such a unique identity can be obtained by using a combination of the address of the moderator, the starting time of the session as measured at the moderator, and a number that is locally unique at the moderator. In many environments, such logs need to be accessible from a single location. In those cases, each peer would need to periodically upload the collected logs to a well-known repository within the system. The unique identity of the session can be used to eliminate duplicates and keep a single copy of each session at the repository.

The peer-to-peer approach used for instant messaging can be used for other collaboration applications that are similar in design. Instant messaging and instant messaging conferences use text for communication. Similar approaches can be used for audio communications (placing telephone calls over the Internet) as well as video communications. The use of peer-to-peer networks for IP telephony is described in Section 9.3.

9.3 IP TELEPHONY

The telephone is a familiar instrument in most of the world, and it allows people to talk to each other over long distances. The telephone end point provides an interface for people to dial the globally unique phone number of the remote person, an interface to speak, and an interface to listen to words spoken on the remote end. Because most personal computers and laptops provide a speaker and a mike as standard equipment, they can be used as the end points for voice communication, effectively setting up a telephone network over the Internet.

Carrying telephone conversation over an IP network has significant economic advantages for the end user. Most consumers have a fixed rate for Internet access, and most enterprises have a fixed rate for their data lines to the campuses. Routing the telephone calls over the Internet allows telephone calls to other users with data network connectivity at no additional charge. When a call is made to an individual who has a traditional phone line and no access to the data network, telephone charges must be incurred on the portion of the call made on the traditional phone network. However, a long-distance call may be converted into a local phone call depending on where one breaks out from the data network into the telephone network.

Most current Voice over IP (VoIP) systems are based on the client-server architecture. A soft-phone (software that implements phone functions) is present on each client to initiate calls to a server in the network. The server typically would have a directory that maps the dialed phone number to the IP address of the receiver. If the user is on the traditional phone network, then the mapping is done to the IP address of the closest point where a call can be placed to break into the traditional phone network. The server would signal the soft-phone on the receiver and provide it information about the calling party. The two parties would establish a session between themselves and exchange IP packets that encapsulate the voice that the parties are involved in. The server could also initiate a conference among more than two users of the IP telephony system. The two established standards for IP telephony, H.323 and SIP, follow a similar mechanism.

Breaking out of the IP network into the telephone network requires special hardware that can only be provided by special servers within the network. However, the group of nodes that are connected on the IP network can use a peer-to-peer architecture among themselves that allows for unlimited audio communication among themselves without needing a server. The ability to break into the existing public telephone network would require the establishment of servers that do the same. In this book, we look at the case of VoIP communication that occurs purely among the users connected to an IP network.

Each of the users must get a unique identifier, a telephone number or some other string, that identifies the user uniquely. This can be done by means of mechanisms that are analogous to those described for getting a unique distinguished name for an entry in a distributed peer-to-peer directory system. In summary, a user can choose a long string randomly so that the chance of collision with another user is negligible, broadcast its selected names to other peers so as to identify any conflicts, or use a server to keep track of registered unique names. Users can call other users by using their unique identifiers or conference more than one user together by using the unique identifiers.

The process for discovering a user that is being called would be completed by doing a broadcast search on the peer-to-peer overlay or by using the search procedure described for distributed peer-to-peer directory systems. Information about users whose locations have been known recently can be cached so that subsequent calls can be made to them quickly without necessarily doing a broadcast search. Once the called party is located, communication can be established and a phone conversation initiated.

Conference calls can similarly be established as long as each peer has software that can perform the task of multiplexing the various calls. Uncompressed voice has a typical bandwidth of 64 Kbps, and standard compression techniques reduce the required bandwidth to 7.2 Kbps. This implies that a user with standard dial-up of 56 Kbps will have the ability to conference about 7 users, whereas users on cable modems or DSL lines could easily provide a conference call that supports about 50 users (or more depending on their connectivity speed). With the right software, any of the peers can act as a conferencing server to host a conference call among a substantially large set of other peers.

Thus, VoIP can be provided as a service among the users of a peer-to-peer network at no additional cost. At the time of the writing of this book, some peer-to-peer systems such as Kazaa had announced the addition of VoIP services to the suite of software that they supported. A peer-to-peer phone company (http://www.phonebazooka.com/) also provides access from its system to the traditional phone network by signing up members who are willing to allow local nontoll calls from their home lines on behalf of other users.

A similar service can also be provided for videoconferencing among users of a peer-to-peer network if the peers have the right hardware and software at their computers. Videoconferencing requires the presence of a camera at each of the participating peers. Because video transmission rates are significantly higher than voice rates, only peers with broadband access (cable modem, DSL, or better) would be able to participate in such a conference. Also, the amount of bandwidth required to perform the conferencing function at a peer would restrict the size of members that can participate in a conference. Thus videoconferencing as a peer-to-peer application would be limited by the current access bandwidth limitations. However, as access bandwidth increases in the future, and video cameras become common peripherals in end devices, one would expect to see increased usage of peer-to-peer videoconferencing as well.

9.4 SHARED COLLABORATION DATABASES

One of the most useful collaboration applications is the use of a shared database where participants can contain files and other documentation related to a common subject. The material contained in the subject matter could consist of reference documents about a common topic, threads of discussions on mail, or progressive copies of work in progress. The collaboration databases allow for different people to access the documents and to comment on them at their own pace.

A server-based implementation of collaboration databases is prevalent in most enterprises and public forums. Web servers that contain threads of E-mail discussions abound on the Internet. A common discussion group without access control restrictions is provided by means of a WIKI [46]. A WIKI is a collaborative open notebook that allows for creation and open editing of shared information by any visitor. It can be viewed as a website whose contents can be modified by any visitor using a standard web -browser. On the enterprise side, collaboration databases as provided by groupware providers (e.g., Lotus Notes or Microsoft Exchange) are used as collaboration tools in different enterprises. It would be useful to have similar functions available on a peer-to-peer infrastructure, without the expense and complexities of operating a server for this purpose.

Such collaboration databases can be created as an application on top of the file storage middleware that was discussed in Chapter 5. File storage middleware provide the ability to store and access files based on a unique identity or handle that was provided to them. Collaboration databases are essentially views that are built on the files that create an index of the different documents that are stored in the repository. This index itself can be considered a special type of file and stored in the distributed peer-to-peer repository that is offered by the file storage service. The recursive use of storing a view as a file is used in file systems based on the Unix file system, and it is a simple extension of that concept to the peer-to-peer paradigm.

To illustrate how peer-to-peer systems can be used to implement collaboration databases, let us examine the development of a WIKI system as a peer-to-peer application. A WIKI system is simple to develop as a website because there are no access control requirements for operating a WIKI. Other collaboration systems enforce stricter access control requirements.

A WIKI is a shared set of pages that can be edited by any of the visitors to the site. Thus each page in the WIKI allows for the ability of a user to edit the contents of the page, add hyperlinks, and create new pages. As a website, this simply implies that links are provided that permit the above operations to be performed on every page. Versions of the pages are maintained after each change, so that any incorrect or malicious updates can be rolled back to a proper version. The order in which requests arrive at the server defines a global snapshot of how the entire WIKI system looks at any given time.

In a peer-to-peer version of a WIKI system, there is no server to arbitrate and maintain the versions of a specific file. However, the file storage service maintains each file on a primary node, the node to which the file handled has been mapped. The file may be replicated to multiple nodes for availability, but at any time each file handled is owned by one primary node. The primary node is responsible for maintaining the versions of the file that it belongs to. Thus requests are serialized at each of the individual pages, but not for an entire set of files. In a WIKI, the lack of such serialization is not a significant issue because the integrity of the information in a file is the primary concern. Because the linkage of information is the primary concern in a WIKI, the set of files are only linked together by the information contained in each individual file. Therefore, ensuring that each file is maintained consistently suffices to address the consistency of the overall information.

The other function provided by a WIKI is the ability to search through all of the information contained in the set of files. This can be implemented as a broadcast search on a peer-to-peer system of files. To ensure that the search only looks for the files that are included within the WIKI information set, the search must ensure that it includes the term that uniquely identifies the WIKI. As long as each of the files included in a WIKI contains a reference to the WIKI to which it belongs, the search would be able to find the information that is pertinent only to that WIKI.

The peer-to-peer implementation of a WIKI can allow the same set of information to belong to more than one WIKI's unique identity. This would enable the sharing of information among different WIKIs in an easier manner than is possible by the implementation of server-based WIKI, where each WIKI server maintains its own information separately. Thus a peer-to-peer implementation of a WIKI may be more advantageous in a community of collaborative users than a server-based approach.

In real life, one may encounter some members of the community who are trying to attack a WIKI in a malicious manner. It may also be desirable to track who is making the modifications in a system. In commercial groupware applications, and in most enterprise collaborative databases, access to the system is controlled by means of access control lists that define who can read, write, or modify a file. In a peer-to-peer file system, such access control can be implemented by means of certificates based on public key cryptography.

Peer-to-peer implementations of collaborative groupware can be used as an alternative to dedicated servers to provide a lower-cost alternative to managing shared information. However, the peer-to-peer implementation would typically tend to be slower in its searches than accessing a server, the availability would likely be worse, and the version control would tend to be more difficult. On the positive side, the sharing of common information among different collaborative databases would be more efficient.

9.5 COLLABORATIVE CONTENT HOSTING

The type of collaborative applications that we have discussed thus far consists of peers trying to maintain shared information. Another type of collaborative support that peer-to-peer systems can offer is to help each other perform operations that none of them is capable of doing alone. This includes the ability to provide a more scalable system than any one of them is capable of hosting itself.

Consider the case of a user who is trying to send out video data from a personal computer located in his basement. Even if the user has broadband access from his site to stream out the video, the available bandwidth is not enough to support more than two or three concurrent clients accessing the system. If the hosting person were collaborating with a set of other peers, each of them capable of supporting two to three concurrent clients, then the set of peers could collaboratively support many more concurrent clients. With 10 peers, each capable of supporting 3 clients, one should be able to support slightly less than 30 concurrent clients accessing the system. The number would be slightly less than 30 because of the overhead of synchronizing among the clients.

One possible way for the collaborative content hosting to be done would be for the primary site to act as a redirector. A large file is replicated at several peers, and each of them is able to serve it to the other clients. The primary site keeps track of the different sites that have a copy of the file being hosted and their capabilities, that is, the number of servers that they can host. Clients contact the primary site to access the file, which can either decide to serve them from the local copy (if there is adequate capacity) or refer the clients to download the file from one of the other sites. The primary site can use a round-robin or random scheme to select the other site so that the load on the different partners is evened out.

The collaboration among the different peers can be used to provide mutual benefits to all of the participating peers. Consider 10 individuals, each of whom has broadband connectivity to the Internet and owns a moderate-sized personal computer at his or her home. Each of them could run a website with a limited scalability. For an illustrative purpose, let us say that each is capable of running a site with a maximum outbound bandwidth of 500 Kbps and a maximum of 10 concurrent clients. If they do not cooperate, each of them is limited by these restrictions. Collectively, all 10 of them are limited to a bandwidth of 5000 Kbps and 100 concurrent clients overall. However, it is highly unlikely that all of their sites are being accessed concurrently at the maximum limit. One of them may be getting a lot of clients in the morning and very few hits in the evening, whereas the situation may be reversed for another one of them. Sharing the excess capacity from one of the peers would help all of them to provide a site with better scalability.

The redirection of peers to other sites may be done in a variety of ways and using different heuristics for balancing the load among them. The peers may exchange their current load with each other and use that information to determine where clients ought to be redirected. The peers could also decide to select one peer as a primary site that collects information about all of them and is used as the initial contact point for the clients. The primary site then directs the clients to one of the other sites so that they can obtain the content from one of the other sites that are available.

One of the mechanisms for different sites to cooperate together would be a domain name server that provides the IP addresses for each of them. Each of the sites designates the domain name server as the one with the ownership of its domain. The domain name server knows where different copies are replicated and redirects the clients to the right machine that can serve a specific contact. Domain name services are only used to resolve machine names;

therefore, the above mechanism works to foster cooperation among sites that are using different machine names as part of their URL.

The peer-to-peer mechanisms described above have been used to develop collaborative web servers in systems such as those described in [47] and [48]. Peer-to-peer collaboration can also be used among web clients to provide for mutually beneficial functions such as the one described in Section 9.6.

9.6 ANONYMOUS WEB SURFING

In the normal course of operations, information about the websites I visit with standard web browser software is accessible to many entities. The Internet service provider I use to access the Internet can track each of the accesses I make, my employer can track all of the sites I visit at work, and so can any person who is able to put a packet sniffer in the network and record the packets going to/from my computer's IP address. In many cases, I do not care that my accesses can be tracked and viewed by anybody. However, there are many instances where I would not like anyone to be tracking information about the sites I visit. Someone may use that access pattern to target me to send junk electronic mail. Users residing in some nondemocratic countries may have valid security concerns for hiding their identity when accessing information that may be considered politically incorrect.

Identity of visitors is also tracked by many of the websites on the Internet with a variety of mechanisms by placing cookies on the browsers or by requiring registration of users with personal information such as E-mail addresses and age. Although most of the sites are trustworthy, some of the sites may still target people for sending junk mail. Some other sites may expose sensitive information if their security is compromised. Some sites may profile my access pattern, and this information can be used incorrectly against me in many cases. As an example, if I access information about speeding tickets on an Internet site, my automobile insurance company, which shares profiles from that site, may incorrectly classify me as a potential speeder and increase my insurance premium, even if I am only researching that subject to write an article in a local magazine.

Such risks also exist in systems that track consumer profiles, for example, in the discount cards offered by many grocery store chains. The grocery chains provide discounts to the owners of those cards, but information collected by the stores could potentially be used to track usage patterns and behavior of the card holder. From a consumer perspective, it would be beneficial to have the discounts while not losing the privacy. In a similar manner, information collected by a website is valuable, although the loss of privacy is an undesirable price to pay for the privilege of accessing those data.

Anecdotal evidence has it that computer science students at many universities have figured out an effective way of maintaining their privacy while retaining the discounts associated with the grocery chain cards. Once a fortnight, a group of students get together for a party, toss all of their grocery store cards into a big hat at the beginning of the party, and then randomly pick one of the cards when they leave the party. The usage pattern for any given card is thus randomized over the group of students and does not reveal any information about any of the students from the sample.

A similar randomization scheme is often used to obtain anonymity by users when surfing the Internet. Clients access their websites through special sites that randomize the cookies assigned by the website to the various users, thereby mixing the access patterns obtained by all of the users. Neither the site, nor anyone observing the traffic out of the anonymizing site, would be able to track the sites that are being accessed by any specific user. Thus anonymizing sites are effective at combining the behaviors of many users into one.

Because anonymizing sites are well-known locations, they still do not solve the problem of anonymity completely. A dictatorial government, for example, could block access to all of the anonymizing sites from users within their domain. This effectively cuts off all Internet access to any user who wishes to preserve his/her anonymity or to access content on any forbidden site. A peer-to-peer approach to anonymity helps to bypass the limitation of a known anonymizing site.

If multiple users come together in a peer-to-peer application in which collaborate with each other, they can provide anonymity to all of themselves in an easy manner. To access a website, a peer does not access it directly but passes the requests on to one of the many different peers. Each peer randomly decides whether it wants to pass the request onto another peer or to obtain the contents from the site. The random decision may be made by the peer generating a pseudorandom number and accessing the website if the generated random number exceeds a specific threshold. The peer obtaining the information from the website passes it back to the neighbor that requested the message, until it eventually reaches the original peer requesting the content.

With a large peer-to-peer group operating behind the firewalls of a nondemocratic country, the behavior of users accessing websites can be effectively randomized so that the access pattern of any single user would be difficult to track. If there are a large number of peers operating outside of the firewalls, the peers can effectively access content from any site (including prohibited ones) while minimizing the probability that the original requestor of the site can be tracked down. Short of prohibiting the operation of any peer-to-peer system, it would be very difficult to isolate the files being accessed by an individual user in such systems.

10

RELATED TOPICS

In this book, we have looked at several useful middleware systems and applications that can be developed by using peer-to-peer systems. We have seen the usage of peer-to-peer systems in different contexts. In this final chapter, we look at some of the topic areas that are closely related to the concept of peer-to-peer computing.

The rise of peer-to-peer computing is generally associated with the growth in popularity of the Napster file sharing program. However, there are many examples of peer-to-peer systems that existed long before Napster was incorporated. The first topic we examine in this chapter is an overview of some of the peer-to-peer applications that existed before Napster made "peer-to-peer" a household name.

Peer-to-peer computing is a mechanism for building distributed applications. Grid computing, another mechanism for building distributed applications, has many similarities to peer-to-peer computing. The second topic that we examine in this chapter is the concept of grid computing and how it relates to peer-to-peer computing.

Legitimate Applications of Peer-to-Peer Networks, by Dinesh C. Verma
ISBN 0-471-46369-8

10.1 LEGACY PEER-TO-PEER APPLICATIONS

At least two very successful peer-to-peer applications were used widely in the Internet before the advent of file sharing programs like Napster. One of these applications is responsible for the operation of the Internet itself and is used for maintaining the information needed for routing packets from one machine to another.

To ensure that packets sent by one machine reach other machines in the Internet, each computer maintains a set of routing tables. The routing tables are computed automatically by the different routers in the Internet. The protocols for computing the routes within the Internet are an instance of a peer-to-peer application that has proven itself within the Internet.

The Internet is a conglomeration of several independent networks run by different companies, each independently administered network segment interacting with many other network segments in a mesh topology. For the purposes of routing, the Internet is divided into several autonomous systems (ASs), each AS consisting of a set of routers under a single administrator's control. The same administrator can control more than one AS. Each AS interconnects with one or more ASs at different points. At each such interconnection point, the routers on each side of the interconnection run an exterior gateway protocol ("gateway" is an alternate name for router) between themselves to exchange routing information between themselves. The routers within an AS run an interior gateway protocol to exchange routing information among themselves. The most common exterior gateway protocol is the Border Gateway Protocol (BGP) [49].

The two BGP daemons on each side of the interconnection exchange information with each other about the IP subnets they can reach on their own sides of the interconnection. For each subnet that can be reached, the BGP routers exchange information about the sequence of AS that must be traversed to reach that subnet. All of the BGP routers that are in the same AS exchange the information obtained from their peers in other ASs with each other. Thus each BGP router will obtain routing information from other routers in the same AS, as well as one (and possibly more than one) router in another AS. The routing information is then combined by each of the BGP routers to determine the set of AS that provides the shortest route (in AS hop count) to each of the subnets about which the route information is known.

The BGP protocol works with each BGP router being a peer to other BGP routers. Although the routers must be configured with each other's identities, they all perform identical and symmetric roles in the execution of the gateway protocol. As such, BGP can be looked upon as one of the first successful peer-to-peer applications to be deployed widely on the Internet.

The interior gateway protocols also work by treating all of the routers within an AS as peers. Most of the interior gateway protocols start with a discovery mechanism, in which a router discovers the identity and IP address of the router that is connecting to it on the other physical link. The subsequent exchange of routing information depends on the routing protocol that is used. One type of routing protocol is the distance vector routing protocol. In these protocols, each router maintains a table with the shortest known lengths to all destinations. Adjacent routers exchange the tables with each other periodically and update their routing tables on the basis of the exchanged information. Examples of distance vector routing protocols include RIP, a protocol used in the Internet since the early 1970s and thus one of the earliest examples of a peer-to-peer application.

Another type of routing protocol is classified as link-state routing protocol. These protocols build the topology of an entire AS at each router and use that topology to compute routing paths. The topology is built by having each router broadcast the status of its links to all of the other routers within the AS. OSPF [13] is an example of a link-state routing protocol.

Another legacy peer-to-peer application successfully deployed within the Internet is the Usenet news system [50]. The Usenet news system consisted of client news readers, which provided the software used to post or read news articles, and news servers, which kept track of messages sent within the group. The news servers implemented a peer-to-peer protocol—NNTP, or the network news transfer protocol [51]—among themselves to circulate any articles posted by a user to all of them.

In the protocol used between the news servers, all the news servers are connected in a mesh, with each news server being connected to one or more of the other news servers. New articles are propagated by a modified form of flooding. At periodic intervals, each news server checks with its neighboring news servers to check whether a new article has arrived. If so, the new article is downloaded from the server if it does not already exist. Messages are prevented from looping indefinitely by having each message carry an identifier, as well as by maintaining a maximum hop count in the message. Readers will readily identify this protocol as similar to the mechanism used in many current peer-to-peer systems used for broadcasting queries to all of the participants.

BGP, OSPF, RIP, and NNTP are all examples of peer-to-peer protocols that were used successfully in the Internet long before anyone had heard of file sharing with Napster or Kazaa.

10.2 GRID COMPUTING

The execution of many scientific applications requires a large amount of processing power and storage space. The computation power needed for such applications required an infrastructure of expensive supercomputers, which could only be established at a few selected supercomputing centers. Each of these supercomputing centers would provide facilities for physicists, chemists, and biologists to rent computing cycles as needed so that they could run their expensive simulations of astronomical objects, molecular biology, or genetic mutations. It did not take long for scientists at different supercomputing centers to realize that they could better use each other's facilities by linking their supercomputing centers together in a grid where computing resources could be borrowed from each other's sites to meet the demands of their users. The borrowing of the computing cycles from disparate sites among a large number of supercomputing sites led to the growth of grid computing.

Grid computing refers to the techniques of hooking together several sites with available computing resources into a seamless assembly of computing resources from which virtual groups can be carved out. These virtual groups of resources could consist of machines from many different locations but are working together to run one common application. The infrastructure that is required to support this type of collaboration among the different computing sites is the grid computing middleware. Some examples of grid computing middleware are the Globus toolkit [52], Avaki [53], and Unicore [54]. There is considerable interest in adopting the grid computing paradigm to commercial applications, resulting in a strong support from companies like IBM to grid standards like OGSA [55].

Grid computing has many similarities to peer-to-peer computing. The different sites in a grid are peers to each other, and the grid middleware needs to implement techniques to discover and share resources present at various sites. At a conceptual level, applications written over grid systems would be similar to applications written over peer-to-peer computing systems, and grid middleware would implement mechanisms that would be similar to those described in peer-to-peer systems in the various chapters of this book.

The difference between grid computing and peer-to-peer computing is in the design point selected for the design of the middleware and applications. Grid computing systems would tend to be designed to operate well for the case in which there are a small number of very powerful locations, whereas peer-to-peer computing systems tend to be designed for a very large number of weaker machines. The difference in operating assumptions leads to emphasis on different aspects of the middleware. Whereas grid computing tends to focus on aspects such as scheduling of jobs within the different resources, peer-to-peer system middleware tends to focus more on self-configuration and recovery from the failure of unreliable machines in the system.

REFERENCES

[1] Internet Assigned Numbers Authority Online Database of Assigned Port Numbers, available at URL HYPERLINK "http://www.iana.org/assignments/port-numbers" *http://www.iana.org/assignments/port-numbers.*

[2] I. Stoica, R. Morris, D. Karger, M. F. Kaashoek, and H. Balakrishnan, *Chord: A scalable peer-to-peer lookup service for Internet applications,* Proceedings of the 2001 Conference on Applications, Technologies, Architectures, and Protocols for Computer Communications, San Diego, CA, pp. 149–160.

[3] S. Ratnasamy et al., *A scalable content-addressable network,* Proceedings of ACM SIGCOMM, San Diego, CA, August 2001.

[4] J. Kubiatowicz et al., *OceanStore: An architecture for global-scale persistent storage,* Proceeedings of the Ninth International Conference on Architectural Support for Programming Languages and Operating Systems (ASPLOS 2000) Boston, MA, November 2000.

[5] Kazaa, URL *http://www.kazaa.com.*

[6] Project JXTA, *http://www.jxta.org.*

[7] Microsoft Windows XP peer-to-peer SDK, *http://msdn.microsoft.com/downloads/list/winxppeer.asp.*

[8] D.Verma, *Content Distribution Networks,* ISBN 0-471-44341-7, Chapter 2, "Site Design and Scalability Issues in Content Distribution Networks," John Wiley & Sons, New York, NY, Dec. 2001.

[9] The Gnu HTTP Tunnel Protocol. Specifications available at *http://www.gnu.org/software/httptunnel/httptunnel.html.*

[10] S. Banerjee, C. Kommareddy, and B. Bhattacharjee, *Scalable Peer Finding on the Internet,* Proceedings of Global Internet Symposium, IEEE GLOBECOM 2002, Taipei, Taiwan, Nov. 2002.

Legitimate Applications of Peer-to-Peer Networks, by Dinesh C. Verma
ISBN 0-471-46369-8

[11] C. Miller, *Multicast Networking & Applications,* ISBN 0-201-30970-3, Addison-Wesley, Oct. 1998.

[12] M. Shin, Y. Kim, K. Park, and S. Kim, Explicit Multicast Extension for Efficient Multicast Packet Delivery, *Korean Electronics and Telecommunications Research Institute Journal,* Vol. 23, No. 4, Dec. 2001. *http://etrij.etri.re.kr/etrij/pdfdata/23-04-L02.pdf.*

[13] J. Moy, *Multicast Extensions to OSPF,* Internet RFC 1584, March 1994. *http://www.ietf.org/rfc/rfc1584.txt.*

[14] T. Bates, R. Chandra, D. Katz, and Y. Rekhter, *Multiprotocol Extensions for BGP-4,* IETF RFC 2283,February 1998, *http://www.ietf.org/rfc/rfc2283.txt.*

[15] IETF Working Group on Reliable Multicast Transport Protocol. Charter available at *http://www.ietf.org/html.charters/rmt-charter.html.*

[16] C. Diot, B. Levine, J. Lyles, H. Kassem, and D. Balensiefen, Deployment issues for the IP multicast service and architecture, *IEEE Network,* Jan 2000, pp. 78–88.

[17] P. Francis, Your Own Internet Distribution, Project description at URL *http://www.aciri.org/yoid/.*

[18] D. Pendarakis, S. Shi, D. Verma, and M. Waldvogel, *ALMI: An application level multicast infrastructure.* Proceedings of the 3rd USENIX Symposium on Internet Technologies and Systems (USITS), 2001, pp. 49–60.

[19] Y. Chu, S. G. Rao, and H. Zhang, *A Case for End System Multicast,* Proceedings of ACM SIGMETRICS 2000.

[20] J. Jannotti, D. Gifford, K. Johnson, M. Kaashoek, and J. O'Toole Jr, *Overcast: Reliable multicasting with an overlay network,* Proceedings of the Fourth Symposium on Operating System Design and Implementation (OSDI), Oct. 2000.

[21] D. Verma et al, SRIRAM: A scalable resilient autonomic mesh, *IBM Systems Journal,* vol. 42, no 1, 2003. *http://www.research.ibm.com/journal/sj/421/verma.pdf.*

[22] D. Goldschlag, M. Reed, and P. Syverson, Onion routing for anonymous and private internet connections, *Communications of the ACM,* vol. 42, no. 2, 1999.

[23] The Open Digital Rights Language Initiative, *http://www.odrl.net.*

[24] Extensible Rights Markup Language, *http://www.xrml.org.*

[25] Secure Digital Music Initiative, *http://www.sdmi.org.*

[26] Interoperability of Data in eCommerce Systems, *http://www.indecs.org.*

[27] Morpheus File Sharing Software, *http://www.morpheus.com/.*

[28] Gnutella, *http://gnutella.wego.com/*

[29] SHA-1 Hash Algorithm, *http://www.itl.nist.gov/fipspubs/fip180-1.htm*

[30] C. Plaxton, R. Rajaraman, and A. Richa, *Accessing nearby copies of replicated objects in a distributed environment.* Proceedings of the ACM SPAA, Newport, RI, June 1997.

[31] A. Rowstron and P. Druschel, *Pastry: Scalable, distributed object location and routing for large-scale peer-to-peer systems,* Middleware, Lecture Notes in Computer Science, vol. 2218, 2001.

[32] G. Malkin, *RIP Version 2,* IETF RFC 2453, Bay Networks, Nov. 1998, *http://www.ietf.org/rfc/rfc2453.txt.*

[33] L. Rizzo, Effective erasure codes for reliable computer communication protocols, *ACM Computer Communication Review,* vol. 27, pp. 24–36, Apr. 1997.

[34] I. Clarke, O. Sandberg, B. Wiley, and T. Hong, *Freenet: A distributed anonymous information storage and retrieval system.* Proceedings of the ICSI Workshop on Design Issues in Anonymity and Unobservability, Berkeley, CA, June 2000.

[35] Veritas Software, *http://www.veritas.com/products/listing/ProductListing.jhtml*

[36] IBM Tivoli Storage Manager, *http://www.ibm.com/software/tivoli/products/storage-mgr/.*

[37] E. Clark, Emerging Technology, Keeping Storage Costs Under Control, *Network Magazine,* Oct. 2002, *http://www.networkmagazine.com/article/NMG20020930S0004.*

[38] B. Mellish, P. Chudasma, S. Lukas, and M. Rosichini, *IBM Enterprise Total Storage Server, IBM Redbook* SG-24-5757, http://www.redbooks.ibm.com/pubs/pdfs/redbooks/sg245757.pdf.

[39] B. Loo and A. LaMarca, *Peer-To-Peer Backup for Personal Area Networks,* Intel Research Technical Report IRS-TR-02-015, May 2003, *http://www.intel-research.net/Publications/Seattle/052820031647_102.pdf.*

[40] M. Wahl, T. Howes, and S. Kille, Lightweight Directory Access Protocol (V3), IETF RFC 2251, Dec. 1997, *http://www.ietf.org/rfc/rfc2251.txt.*

[41] LDAP Duplication/Replication/Update Protocols (ldup), IETF Working Group. Charter available at *http://www.ietf.org/html.charters/ldup-charter.html.*

[42] T. Howes, M. Smith and G. Good, Understanding and deploying *LDAP Directory Services,* ISBN 0-672-32316-8, Addison-Wesley, May 2003.

[43] M. Rizcallah, *LDAP Directories,* ISBN 0-470-84388-8, John Wiley & Sons, Hoboken, NJ, Nov. 2003.

[44] R. van Renesse, K. Birman, and S. Maffeis, Horus, A Flexible Group Communication System, *Communications of the ACM,* April 1996.

[45] Y. Al-Houmaily and P. Chrysanthis, *Two-Phase Commit in Gigabit-Networked Distributed Databases,* Proceedings of 8th Intl. Conf. on Parallel and Distributed Computing Systems, Sept. 1995.

[46] B. Leuf and W. Cunningham, *The Wiki Way: Quick Collaboration on the Web,* ISBN 020171499X, Addison-Wesley, Apr. 2001.

[47] R. Bayardo Jr., A. Somani, D. Gruhl, and R. Agrawal, *YouServ: A Web Hosting and Content Sharing Tool for the Masses,* Proceedings of the 11th Intl World Wide Web Conference (WWW-2002), 2002.

[48] S. Jin and A. Bestavros, *Cache-and-Relay Streaming Media Delivery for Asynchronous Clients,* Proceedings of Networked Group Communication, Boston, MA, 2002.

[49] S. Thomas, *IP Switching and Routing Essentials: Understanding RIP, OSPF, BGP, MPLS, CR-LDP and RSVP-TE,* ISBN 0-471-03466-5, John Wiley & Sons, New York, NY, Dec. 2001.

[50] M. Hauben, R. Hauben, and T. Truscott, *Netizens: On the History and Impact of Usenet and the Internet,* ISBN 0-818-67706-6, IEEE Computer Society Press, Apr. 1997.

[51] B. Kantor and P. Lapsley, *Network News Transfer Protocol,* IETF RFC 977, Feb. 1986, *http://www.ietf.org/rfc/rfc977.txt.*

[52] The Globus Alliance, *http://www.globus.org.*

[53] Avaki Data Grid, available at http://www.avali.com.

[54] UnicorePro Grid Software, available at *http://www.pallas.com/e/products/unicore-pro/index.htm.*

[55] Open Grid Services Architecture, Specifications available at *http://www.globus.org/ogsa/.*

INDEX

Legitimate Applications of Peer-to-Peer Networks, by Dinesh C. Verma
ISBN 0-471-46369-8